Data Analysis in High Energy Physics

Data Analysis in High Energy Physics

Contributors

Andrew Walcott Beckwith et al.

AURIS
Reference

www.aurisreference.com

Data Analysis in High Energy Physics

Contributors: Andrew Walcott Beckwith et al.

Published by Auris Reference Limited

www.aurisreference.com

United Kingdom

Data Analysis in High Energy Physics

ISBN: 978-1-78154-886-8

British Library Cataloguing in Publication Data
A CIP record for this book is available from the British Library

Printed in the United Kingdom .

Exclusively distributed by CBS Publishers & Distributors Pvt. Ltd.

Sales & Distribution Rights only for India, Pakistan, Bangladesh, Sri Lanka, Nepal and Bhutan.This book is not to be sold outside these territories.

Contents

List of Abbreviations

AMC	Advance Mezzanine Cards
AOD	Analysis Object Data
APDs	Avalanche Photodiodes
BES	Beijing Spectrometer
CPU	Central Processing Unit
CAS	Chinese Academy of Sciences
CMS	Compact Muon Solenoid
CPM	Critical Path Method
DAQ	Data acquisition system
DSP	Digital Signal Processing
DSN	Distributed sensor network
ENOB	Effective number of bits
EMC	Electromagnetic Calorimeter
EMI	European Middleware Initiative
ESD	Event Summary Data
GUID	Globally Unique Identifiers
GPU	Graphics Processing Unit
HEP	High Energy Physics
HVP	High voltage platform
IHIP	Institute of Heavy Ion Physics
ICs	Integrated Circuits
IGM	Intergalactic medium
KE	Kinetic energy
LHC	Large Hadron Collider
LFN	Logical File Names
LTDA	Long Term Data Access
LEBT	Low energy beam transport
MDC	Main drift chamber
MWS	Microwave Studio
MAC	Multiply-Accumulate
MUC	Muon identifier chamber
NDGF	Nordic Datagrid Facility
NPP	Nuclear Power Plant
OS	Operating system
OPN	Optical Private Network
PDS	Permanent data storage
PC	Personal Computer
PMT	Photomultiplier Tubes
PFN	Physical File Names
PSA	Pulse height analysis
QGP	Quark Gluon Plasma

RFQ	Radio frequency quadrupole
RTM	Rear Transition Modules
XRR	RX reflectometry
SNR	Signal to Noise
SGM	Software Group Manager
SR	Special relativity
SLC	Spiral Loaded Cavity
SDTY	Standard data taking year
TAC	Time-to-amplitude converter
TDRs	Technical Design Reports

List of Contributors

Andrew Walcott Beckwith
Physics Department, Chongqing University Huxi Campus, Chongqing, China

D M South
Deutsches Elektronen Synchrotron, Notkestraße 85, 22607 Hamburg, Germany

V. González
University of Valencia, Dep. of Electronic Engineering, Spain

D. Barrientos
University of Valencia, Dep. of Electronic Engineering, Spain
Instituto de Física Corpuscular (IFIC), Valencia, Spain

J. M. Blasco
University of Valencia, Dep. of Electronic Engineering, Spain

F. Carrió
University of Valencia, Dep. of Electronic Engineering, Spain
Instituto de Física Corpuscular (IFIC), Valencia, Spain

X. Egea
University of Valencia, Dep. of Electronic Engineering, Spain

E. Sanchis
University of Valencia, Dep. of Electronic Engineering, Spain

M. S. El Naschie
Department of Physics, Faculty of Science, University of Alexandria, Alexandria, Egypt

S. Olsen
College of Central Florida, Ocala, USA

J. H. He
Nantong Textile Institute, National Engineering Laboratory for Modern Silk, College of Textile and Clothing, Soochow University, Suzhou, China

S. Nada
Mathematics Department, Qatar University, Doha, Qatar

L. Marek-Crnjac
Technical School Center of Maribor, Maribor, Slovenia

A. Helal
Department of Mathematics, Faculty of Science, Cairo University, Giza, Egypt

Xuemin Zhu
The Institute of High Energy Physics, Chinese Academy of Sciences, China

Sen Qian
The Institute of High Energy Physics, Chinese Academy of Sciences, China

Dagmar Adamová
Nuclear Physics Institute, Rež near Prague ˇ Czech Republic

Pablo Saiz
CERN Switzerland

Daniele Rinaldi
Università Politecnica delle Marche/ Dipartimento di Scienze e Ingegneria della Materia, dell'Ambiente e Urbanistica, Switzerland, Italy

Michel Lebeau
European Organization for Nuclear Research (CERN)/Department PH

Nicola Paone
Università Politecnica delle Marche/ Dipartimento di Ingegneria Industriale e Scienze Matematiche

Lorenzo Scalise
Università Politecnica delle Marche/ Dipartimento di Ingegneria Industriale e Scienze Matematiche

Paolo Pietroni (formerly with[3])
Università Politecnica delle Marche/ Dipartimento di Ingegneria Industriale e Scienze Matematiche

Yuancun Nie
State Nuclear Power Technology R&D Center, Beijing
State Key Lab of Nuclear Physics and Technology (Peking University), Beijing China

Yuanrong Lu
State Key Lab of Nuclear Physics and Technology (Peking University), Beijing China

Xueqing Yan
State Key Lab of Nuclear Physics and Technology (Peking University), Beijing China

Jiaer Chen
State Key Lab of Nuclear Physics and Technology (Peking University), Beijing China

Andrew Walcott Beckwith
College of Physics, Chongqing University Huxi Campus, Chongqing, China

Lorenzo Zaninetti
Physics Department, Turin, Italy

Bharat Singh
Department of Information Technology, Indian Institute of Information Technology, Allahabad, India

Nidhi Kushwaha
Department of Information Technology, Indian Institute of Information Technology, Allahabad, India

Om Prakash Vyas
Department of Information Technology, Indian Institute of Information Technology, Allahabad, India

Preface

Data Analysis in High Energy Physics covers the essential tasks in statistical data analysis encountered in high energy physics and provides comprehensive advice for typical questions and problems. The basic methods for inferring results from data are presented as well as tools for advanced tasks such as improving the signal-to-background ratio, correcting detector effects, determining systematics and many others First chapter provides a mechanism for the dominance of kinetic energy in pre- Planckian space-time, as well as its reversal in the Planckian era of cosmology. Second chapter expands on the ideas contained in the original publication and provides a more solid set of recommendations, not only concerning data preservation and its implementation in high-energy physics, but also the future direction and organisational model of the study group. Third chapter presents an overview of technological aspects related to data acquisition (DAQ) systems for particle physics experiments. Being a general topic as data acquisition can be, particle physics experiments pose some challenges which deserve a special description and for which special solutions may be adopted. In fourth chapter, transfinite fractal sets and fuzzy logic are combined to enable the introduction of a new theory termed fractal logic to the foundation of high energy particle physics. Fifth chapter focuses on the quality management of the R&D in high energy physics detector. In sixth chapter, we give a short overview of the Grid computing for the experiments at the LHC and the basics of the mission of the WLCG. Seventh chapter focuses on quality control and characterization of scintillating crystals for high energy physics and medical applications. In eighth chapter, issues of applying the radio frequency quadrupole (RFQ) accelerator to ion implantation for material modification and microelectronics will be discussed at first. Gedanken experiment for refining the unruh metric tensor uncertainty principle via schwartz shield geometry and planckian space-time with initial nonzero entropy and applying the riemannian-penrose inequality and initial kinetic energy for a lower bound to graviton mass is presented in ninth chapter. In last chapter, an analytical law for the evolution of the magnetic field along the radio-jets is deduced using a linear relation between the magnetic pressure and the rest density.

Chapter 1

GEDANKEN EXPERIMENT EXAMINING HOW KINETIC ENERGY WOULD DOMINATE POTENTIAL ENERGY, IN PRE-PLANCKIAN SPACE-TIME PHYSICS, AND ALLOW US TO AVOID THE BICEP 2 MISTAKE

Andrew Walcott Beckwith

Physics Department, Chongqing University Huxi Campus, Chongqing, China

ABSTRACT

We use Padmabhan's "Invitation to Astrophysics" formalism of a scalar field evolution of the early universe, from first principles, to show something which seems counter intuitive. How could, just before inflation, kinetic energy be larger than potential energy in pre-Planckian physics, and what physics mechanism is responsible for the Planckian physics result that Potential energy is far larger than kinetic energy. This document answers that question, as well as provides a mechanism for the dominance of kinetic energy in pre-Planckian space-time, as well as its reversal in the Planckian era of cosmology.

The kinetic energy is proportional to $\rho_w \sim g^* T^4$, with g^* initial degrees of freedom, and T the initial temperature just before the onset of inflation. Our key assumption is the smallness of curvature, as given in the first equation, which permits adoption of the Potential energy and Kinetic energy formalism used, in the Planckian and pre-Planckian space-time physics. Interpretation of this result, if done correctly, will be able to allow a correct distinguishing of relic gravitational waves, as to avoid the BICEP 2 pickup of galactic dust as a false relic Gravitational wave signal, as well as serve as an investigative template as to if quantum gravity is embedded in a deterministic dissipative system, as cited in the conclusion.

INTRODUCTION

We begin with a review from T. Padmanabhan [1] as to the foundations of a scalar field and a potential field, in terms of cosmological evolution. Following that, we are adding more detail as to a supposition by Handley et al. [2] . As to how one could invert the supposition of inflation, [1] , that the Kinetic energy would be much larger than the Potential energy [2] . Here, we offer a mechanism for how this may happen. This is in lieu as to C. VAN DEN BROECK in [3] namely that we need to distinguish between multiple early universe sources of gravitational wave, or the onset of inflation generated early gravitational waves, as seen in the below quotation

"Omni-directional gravitational wave background radiation could arise from fundamental processes in the early Universe, or from the superposition of a large number of signals with a point-like origin. Examples of the former include parametric amplification of gravitational vacuum fluctuations during the inflationary era"

In particular our approach is similar to the procedures outlined as to vacuum fluctuations in the (very early) inflationary era. And we will outline how we are avoiding multiple sources. Which would be done if there is a sharply focused peak frequency in magnitude and intensity, i.e. we wager that a sharply defined peak of GW amplitude and strain is about the only way to avoid the problems associated with Bicep 2, which was due to dust from galactic sources [3] - [5] .

Our gedanken experiment is a thought experiment, that kinetic energy dominates potential energy in the initial phase of cosmological evolution. As a second component of our thought experiment is then the smallness of curvature, so to a good first approximation so that then the formulation of near flatness in the beginning of space-time is a starting point for surprisingly traditional looking Friedman cosmological equations, due to the minimal influence of curvature, which would be a consequence of [6] .

Furthermore in our thought experiment is the supposition that the dominance of kinetic energy is in the regime of space-time before a Planck time interval has transpired. Implicit in this is the supposition that there is such a regime of space-time which the author has also worked out in [7] . If there is a regime of space-time just before Planck time, and this convention adheres to [2] that kinetic energy is, indeed greater than potential energy. In the pre-Planckian space time regime we will have, then a kinetic energy term with imaginary time, and that then the points just before Planck time will have a Potential energy as given by an imaginary time component as set by

$$\dot{\phi}^2 \left(\tau_{\text{Pre-Planckian-time}} \right) \Big|_{\tau_{\text{Pre-Planckian-time}} = \frac{\pm i \times 10^{-44} \, s}{10^{-44} \, s}} \tag{1}$$

Here, what has done, is to scale real to imaginary time, and to normalize it by a division of 10^{-44} seconds, effectively having a "rescaled" pre-Planckian time which would be written as an imaginary number with

$$\tau_{\text{Pre-Planckian-time}} = \frac{\pm i \times 10^{-44} \text{ s}}{10^{-44} \text{ s}}$$

(2)

This means that in an interval just before the Planck time, that we are postulating a pre-Planckian space-time which is purely imaginary, i.e. operationally Equation (2) is, effectively in the regime of analysis of Kinetic energy dominance given by

$$\tau_{\text{Pre-Planckian-time}} = \pm i$$

(3)

The change in time from imaginary to real time, is in line with a full evaluation of a transition from Equation (3) to the regime of space-time dominated by

$$t_{\text{real-time}} = \frac{t_{\text{measured-time}}}{t_{\text{Planck}} \sim 10^{-44} \text{ s}}$$

(4)

In particular what we are supposing is a transformation from the pre-Planckian to Planckian regime given by

$$\tau_{\text{Pre-Planckian-time}} = \left(\frac{\pm i \times 10^{-44} \text{ s}}{10^{-44} \text{ s}}\right) \xrightarrow{\text{Causal-barrier}} t_{\text{real-time}} = \left(\frac{t_{\text{measured-time}}}{t_{\text{Planck}} \sim 10^{-44} \text{ s}}\right)$$

(5)

The right hand side would represent non imaginary time, whereas the regime where there is imaginary structure would represent sub divisions in causal structure of space-time, as alluded to by Dowker [5] [8] with what is represented in Equation (5) as a division from pre-Planckian to Planckian space-time, and a defacto Causal structure with the imaginary time component one which is "tunneled" through, to reach Planckian Space-time physics.

What is conserved between the transference from a prior universe, to present universe would be at the Planckian regime, a minimum stress energy Tensor, which by [3] is given as a term with has the label of T_{00} for transferred space-time energy, and then a measure of curvature of space-time as given in [6] as, if

$$T_{00} = \rho_{\text{Energy-density}} = \frac{-(g_{00} = 1)}{16 \cdot \pi} \cdot \left(\frac{3\tilde{k}_{\text{Curvature-measure}}}{a_{\text{initial-scale-factor}}^2} + \Lambda_{\text{initial-value}}\right)$$

(6)

The $\tilde{k}_{\text{Curvature-measure}}$ is defined in [6] and can be usually thought of as how gravity affects space-time geometry, as given in [9] which is part of the definition of a Riemann Scalar, for which as given in [6] we have that the Riemann scalar is given as

$$\Re = \text{Riemann-scalar} = \frac{-6 \cdot \tilde{k}_{\text{Curvature-measure}}}{\left(a_{\text{initial-value}}\right)^2} \quad (7)$$

The term for a minimum scale factor is given in Equation (7) which is the smallest unit of evolving space-time which is accessible to analysis, which is akin to adopting the bounce definition of a non-singular starting point of a universe expansion as given in [7] [8] [10] [11] with a minimum initial scale value given in [3] [6] as

$$a_{\text{initial-value}} \sim 10^{-55} \quad (8)$$

If so, then we find then that, if we are looking at a cosmological constant, as defined in [9] [10] [12] [13] with an initial value of this cosmological constant as written as $\Lambda_{\text{initial-value}}$ (leaving open the possibility that the cosmological constant could have changed over time, i.e. quintessence, as seen in [11] [14]), that then, as given in part by [3] [6] , where $V^{(3)}$ is three space volume, not including the time dimension,

$$T_{00} = \rho_{\text{Energy-density}} = \frac{-(g_{00}=1)}{16 \cdot \pi} \cdot \left(\frac{3\tilde{k}_{\text{Curvature-measure}}}{a^2_{\text{initial-scale-factor}}} + \Lambda_{\text{initial-value}} \right)$$

$$\Leftrightarrow \tilde{k}_{\text{Curvature-measure}} = -\frac{a^2_{\text{initial-scale-factor}}}{3} \times \left[\left(16 \cdot \pi \cdot \left(\rho_{\text{Energy-density}} = \left[\frac{S_{\text{initial-Entropy}} \cdot m_{\text{graviton}}}{V^{(3)}} \right] \right) \right) + \Lambda_{\text{initial-value}} \right] \quad (9)$$

Here in part we are partaking of the idea of a massive graviton, of about 10^{-62} grams, as given by [12] [15] . and we also, to those whom object to the idea of a massive Graviton refer to Maggiore [13] [16]

$$-3m_{\text{graviton}} h = \frac{\kappa}{2} \cdot T \quad (10)$$

Here, the T is the trace of the Einstein Stress energy Tensor, whereas h is the trace of $h_{\mu\nu}$. In our case, if one took the trace of the massless graviton, $h_{\mu\nu}$ would be identically zero in line with T = 0, whereas in the formulation given by Maggiorie [16] if one did not have the trace of $h_{\mu\nu}$ not equal to zero, one is assuming a modification of the usual massless spin 2 graviton, whereas in terms of our treatment of T, we are effectively, due to the work we did in [7] restricting the T to be equivalent to T_{00} in the pre-Planckian treatment of space-time which is equivalent to looking at, if T_{00} is the same as the time component of the Stress energy tensor in the pre-Planckian regime of space time, that we have by [7] that

$$T\left(= traceT_{uv}\right) \approx \Delta T_{u} \sim \Delta\rho \sim \frac{\Delta E}{V^{(3)}}$$

$$\frac{\Delta E}{V^{(3)}} = \frac{h}{\delta t \cdot \delta g_{u} \cdot V^{(3)}} \equiv \frac{h}{\delta t \cdot V^{(3)} \cdot a^{2}(t) \cdot \phi}$$

$$\Leftrightarrow \Delta m_{graviton}^{2}\left(\text{rest-mass}\right) \sim \frac{\left(1 - \left[v(\text{velocity})_{graviton}/c\right]^{2}\right)c^{2}}{72\pi^{2}h} \times \left[\frac{\kappa}{\delta t \cdot V^{(3)} \cdot a^{2}(t) \cdot \phi}\left(\frac{2\pi v(\text{velocity})_{graviton}}{\Delta\omega_{graviton}}\right)\right]^{2}$$

(11)

The term of the ϕ inflaton scalar field factor shows up in the Equation (13) whereas we define the Graviton coupling term κ [16] via, if G is the gravitational constant, we define it as follows, namely

$$G = \frac{4}{3} \cdot \left(\frac{\kappa^{2}}{32\pi}\right)$$

(12)

If the curvature measure, above is almost zero, then we can use from [1] , with $V(\phi)$ a cosmological potential energy term (usually $V(\phi)$ is more dominant in inflation, to the kinetic energy, but our thought experiment has that the kinetic energy we call $\dot{\phi}$, i.e. the time derivative of the inflaton field, is the dominant factor before the Planck time regime which starts about 10^{-44}seconds), and the term we will be watching as extremely significant, $\dot{\phi}$, the time derivative of the inflaton field. Here, we define the dynamical equation for H, below i.e. that

$$H = \frac{\dot{a}}{a} = \sqrt{\frac{8\pi G}{3}} \cdot \left[V(\phi) + \frac{\dot{\phi}}{2}\right]^{1/2}$$

(13)

If so, then using the Potential energy and Kinetic energy values from [1] we can write the following: Keep in mind that in Equation (14) that the H referred to is defined by Equation (13) above, and so we write below explicit entries for the Potential energy, $V(\phi)$, the Kinetic energy $\dot{\phi}$, and we define a value for Equation (13).

$$V(t) = \frac{3H^2}{8\pi G} \cdot \left(1 + \frac{\dot{H}}{3H^2}\right)$$

$$\phi(t) = \int dt \cdot \sqrt{-\frac{\dot{H}}{4\pi G}}$$

$$H = \frac{2}{t \cdot \left(1 + \frac{p}{\rho}\right)} \sim \frac{2 \cdot t_{\text{REAL-TIME}}^{-1}}{\xi^+}$$

$$\Rightarrow \phi(t) \sim \sqrt{\frac{1}{4\pi G}} \cdot \frac{1}{\xi^+} \cdot \log\left[\frac{t_{\text{final-real-time}}}{t_{\text{initial-real-time}}}\right]$$

$$\Rightarrow \dot{\phi}^2(t) \sim \frac{1}{4\pi G} \cdot \left(\frac{1}{\xi^+}\right)^2 \cdot \left(\frac{1}{t_{\text{real-time}}}\right)^2$$

$$\& \; V(t) \sim \frac{3}{2\pi G} \cdot \left(\frac{1}{\xi^+}\right)^2 \cdot \left[\left(\frac{1}{t_{\text{real-time}}}\right)^2 \cdot \left(1 + \left(\frac{1}{t_{\text{real-time}}}\right)^2\right)\right]$$

$$(14)$$

The term $V(t)$ is for Potential energy, and it is by inspection » $\dot{\phi}^2(t)$ in the Planckian space-time regime which is the Kinetic energy component, provided that time here is a real co-ordinate.

We will, for now on, to keep this real time non dimensional, make the following identification with, once again, Equation (4), which has $t_{\text{real-time}} = \frac{t_{\text{measured-time}}}{t_{\text{Planck}} \sim 10^{-44} \text{ s}}$. For the sake of identification, we will be assuming that Equation (14) and Equation (4) are in the present universe and that ξ^+ is extraordinarily small.

RE EXAMINATION OF EQUATION (14) AND EQUATION (4) IN A PRE UNIVERSE CONFIGURATION

Our supposition is that Equation (2) to Equation (5) in the matter of pre-Planckian space time, say in a boundary of 2 times Planck time to buttress the repeating cyclical universe we are assuming as possibility given by Penrose [17] , is changed then to take into the quantum bounce analogy we think should be looked at [11] as given by C. Rovelli and F. Vidotto. So then we get from Planckian space time, a real time evaluation which shrinks to imaginary time, via the following rule

$$t_{\text{real-time}} = \frac{t_{\text{measured-time}}}{t_{\text{Planck}} \sim 10^{-44} \text{ s}} \xrightarrow{\text{Pre-Planckian}} \tau_{\text{Pre-Planckian-time}} = i \cdot \frac{t_{\text{measured-time}}}{t_{\text{Planck}} \sim 10^{-44} \text{ s}} = i \cdot t_{\text{real-time}}$$

$$(15)$$

i.e. what we are saying is that, there is a retime in the pre-Planckian regime of space-time

In a boundary of $\tau_{\text{Pre-Planckian-time}} \sim \dfrac{\pm i \times 10^{-44}\ s}{10^{-44}\ s}$ i.e. about a bounce area, of space time, then there would be this switch, so then in this regime, we would re write the relevant evaluative time for the Potential and Kinetic energy as $\tau_{\text{Pre-Planckian-time}} \sim \dfrac{\pm i \times 10^{-44}\ s}{10^{-44}\ s}$ .Pick the following point of evaluation, namely at the transit point between the plus to the minus regions of $\tau_{\text{Pre-Planckian-time}} \sim \dfrac{\pm i \times 10^{-44}\ s}{10^{-44}\ s}$ that we are looking at a Vanishing Potential energy, but a Kinetic energy which would be very different from Zero.

$$\dot{\phi}^2\left(\tau_{\text{Pre-Planckian-time}}\right)\Big|_{\tau_{\text{Pre-Planckian-time}}=\frac{\pm i \times 10^{-44}\ s}{10^{-44}\ s}} \sim \frac{1}{4\pi G}\cdot\left(\frac{1}{\xi^+}\right)^2\cdot\left(\frac{1}{\tau_{\text{Pre-Planckian-time}}}\right)^2 < 0$$

$$\&\ V\left(\tau_{\text{Pre-Planckian-time}}\right)\Big|_{\tau_{\text{Pre-Planckian-time}}=\frac{\pm i \times 10^{-44}\ s}{10^{-44}\ s}} \sim \frac{3}{2\pi G}\cdot\left(\frac{1}{\xi^+}\right)^2\cdot\left[\left(\frac{1}{\tau_{\text{Pre-Planckian-time}}}\right)^2\cdot\left(1+\left(\frac{1}{\tau_{\text{Pre-Planckian-time}}}\right)^2\right)\right] = 0 \qquad (16)$$

The fact we have a very large non zero $\dot{\phi}^2\left(\tau_{\text{Pre-Planckian-time}}\right)\Big|_{\tau_{\text{Pre-Planckian-time}}=\frac{\pm i \times 10^{-44}\ s}{10^{-44}\ s}}$ going into the $\tau_{\text{Pre-Planckian-time}} \sim \dfrac{\pm i \times 10^{-44}\ s}{10^{-44}\ s}$ region, as a pre-Planckian bounce bubble, with this flipping to $t_{\text{real-time}} = \dfrac{t_{\text{measured-time}}}{t_{\text{Planck}} \sim 10^{-44}\ s}$ with the result that.

$$\dot{\phi}^2\left(\tau_{\text{Pre-Planckian-time}}\right)\Big|_{\tau_{\text{Pre-Planckian-time}}=\frac{\pm i \times 10^{-44}\ s}{10^{-44}\ s}}$$

$$\xrightarrow[\tau_{\text{Pre-Planckian-time}}\to t_{\text{real-time}}]{} V\left(t_{\text{real-time}}\right) \sim \frac{3}{2\pi G}\cdot\left(\frac{1}{\xi^+}\right)^2\cdot\left[\left(\frac{1}{t_{\text{real-time}}}\right)^2\cdot\left(1+\left(\frac{1}{t_{\text{real-time}}}\right)^2\right)\right] \neq 0 \qquad (17)$$

In making this evaluation, we are assuming that there could be use of the following for relic Gravitational waves., i.e. for Equation (17) to hold we will be looking at a time interval which may be specified by [18] [19]

$$\left(\delta t\right)^2_{\text{emergent}} = \frac{\sum_i m_i l_i \cdot l_i}{2\cdot\left(E-V\right)} \to \frac{m_{\text{graviton}} l_P \cdot l_P}{2\cdot\left(E-V\right)} \qquad (18)$$

Initially, as postulated by Babour [18] [19] this set of masses, given in the emergent time structure could be for say the planetary masses of each contribution of the solar system. Our identification is to have an initial mass value, at the start of creation, for an individual graviton. So If $\left(\delta t\right)^2_{\text{emergent}} = \delta t^2 \sim t_{\text{real-time}} = \dfrac{t_{\text{measured-time}}}{t_{\text{Planck}} \sim 10^{-44}\ s}$ Then there may be gravitons which are [18] [19]

$$m_{\text{graviton}} \geq \frac{2\hbar^2}{\left(\delta g_{tt}\right)^2 l_P^2} \cdot \frac{(E-V)}{\Delta T_{tt}^2}$$

(19)

This would entail assuming relic gravitation generated by a massive graviton bounded below by

$$m_{\text{graviton}} \geq \frac{2\hbar^2}{\left(\delta g_{tt}\right)^2 l_P^2} \cdot \frac{(E-V)}{\Delta T_{tt}^2}\Bigg|_{\tau_{\text{Pre-Planckian-time}}=\frac{\pm i \times 10^{-44}\,s}{10^{-44}\,s}} \rightarrow \frac{2\hbar^2}{\left(\delta g_{tt}\right)^2 l_P^2} \cdot \frac{\left|\dot{\phi}^2\left(\tau_{\text{Pre-Planckian-time}}\right)\right|_{\tau_{\text{Pre-Planckian-time}}=\frac{\pm i \times 10^{-44}\,s}{10^{-44}\,s}}}{\Delta T_{tt}^2}$$

(20)

And the magnitude of K.E. as defined by

$$(E-V) \sim \left|\dot{\phi}^2\left(\tau_{\text{Pre-Planckian-time}}\right)\right|_{\tau_{\text{Pre-Planckian-time}}=\frac{\pm i \times 10^{-44}\,s}{10^{-44}\,s}}$$

(21)

If so, then if we use Equation (16) and Equation (20) and Equation (21) so as to obtain

$$m_{\text{graviton}} \geq \frac{2\hbar^2}{\left(\delta g_{tt}\right)^2 l_P^2} \cdot \frac{\left|\dot{\phi}^2\left(\tau_{\text{Pre-Planckian-time}}\right)\right|_{\tau_{\text{Pre-Planckian-time}}=\frac{\pm i \times 10^{-44}\,s}{10^{-44}\,s}} = \frac{1}{4\pi G} \cdot \left(\frac{1}{\xi^+}\right)^2 \cdot \left|\left(\frac{1}{\tau_{\text{Pre-Planckian-time}}}\right)\right|^2}{\Delta T_{tt}^2}$$

(22)

If so, then, we come to a conclusion, which uses a basic energy density result from Kolb and Turner [9] that the Kinetic energy, as defined in pre-Planckian physics, as defined in this document is decisively important, as given in the conclusion.

RE-EXAMINING RELIC GRAVITATIONAL WAVE MODELS AS TO WHAT RELIC GRAVITATIONAL WAVES COULD TELL US ABOUT THE ORIGINS OF THE EARLY UNIVERSE

It is very noticeable that in [21] we have that the following quote is particularly relevant to consider, in lieu of our results

"Thus, if advanced projects on the detection of GWs will improve their sensitivity allowing to perform a GWs astronomy (this is due because signals from GWs are quite weak) [16] , one will only have to look the interferometer response functions to understand if General Relativity is the definitive theory of gravity. In fact, if only the two response functions (2) and (19) will be present, we will conclude that General Relativity is definitive. If the response function (22) will be present too, we will conclude that massless Scalar-Tensor Gravity is the correct theory of gravitation. Finally, if a longitudinal response function will be present, i.e. Equation (25) for a wave propagating parallel to one interferometer arm, or its generalization to angular dependences, we will learn that the correct theory of gravity will be massive Scalar-Tensor Gravity which is equivalent to f(R) theories. In any case, such response functions will represent the definitive test for General Relativity. This is because General Relativity is the only gravity theory which admits only the two response functions (2) and (19) [21] . Such response functions correspond to the two "canonical" polarizations h+ and h×. Thus, if a third polarization will be present, a third response function will be detected by GWs interferometers and this fact will rule out General Relativity like the definitive theory of gravity"

We argue that a third polarization in Gravitational waves from the early universe may be detected, if there is proof positive that in the pre-Planckian regime that the Corda conjecture [20] as given below, namely if the following analysis is part of our take on relic gravitational waves, is supported by the kinetic energy being larger than the potential energy, namely what if

"The case of massless Scalar-Tensor Gravity has been discussed in [22] [23] with a "bouncing photons analysis" similar to the previous one. In this case, the line-element in the TT gauge can be extended with one more polarization, labelled with $\Phi(t + z)$, i.e. ..."

i.e. the dominance of Kinetic energy over Potential energy, in pre-Planckian physics could serve as a template for verification for the existence of a third polarization along the lines brought up in [20] and its confirmation or falsification would yield foundational insight available nowhere else possible.

CONCLUSIONS

Our hypothesis, as to Equation (22), is equivalent to what is frequently postulated as an energy density as given by Kolb and Turner [12] . First of all the below is equivalent to T_{00}, i.e. the T_{00} is the same as

$$\rho_w \propto a^{-3(1-w)} \sim g^* T^4 \tag{23}$$

If so, then the lower bound to the graviton will be then as given by Equation (24) below, if we use Equation (23), then

$$m_{graviton} \geq \frac{2\hbar^2}{(\delta g_{tt})^2 \, l_P^2} \cdot \frac{\frac{1}{4\pi G} \cdot \left(\frac{1}{\xi^+}\right)^2 \cdot \left|\left(\frac{1}{\tau_{\text{Pre-Planckian-time}}}\right)^2\right|}{\Delta T_{tt}^2 = \left(\rho_w \propto a^{-3(1-w)} \sim g^* T^4\right)^2} \tag{24}$$

i.e. if we have a comparatively low initial temperature, T, it will mean a large initial graviton mass. If the T is of the order of Planck temperature, say

10^{32} Kelvin, then the above will have a lower graviton mass value of about 10^{-66} grams. It goes up if there is what is called a (colder cosmology) about 1 order of magnitude lower initial temperature, leading to the mass of a graviton bounded below by 10^{-58} grams. Really cold initial temperatures far lower than 10^{32} degrees Kelvin for T would lead to maybe even 10^{-50} grams for the initial lower bound to the graviton mass.

For consistency with the 10^{-62} gram value as given by [15] we would probably be considering it desirable for 10^{32} degrees Kelvin for T. In all this we will be considering g^* initial degrees of freedom, of about 100, in terms of what was given by Kolb and Turner [12] .

It is worth noting that Dr. Corda in [21] has extended the Maggiorie results [16] as given in the prior reference [21] section and that indeed Maggiore studied the detectability only for GWs having a wavelength very much longer than the interferometer's arms, while Corda [21] extended the results to all the GWs wavelengths. The importance of this contribution is, if we find out if there is a third polarization as indicated above, possibly due to a dominance of kinetic energy, i.e. the dominance in a pre-Planckian mode of space time may allow for settling the question given in [21] , with an appropriately chosen magnitude, and frequency, and also allow for avoiding the mistake of Bicep 2, as given in references [21] [24] .

We also state unequivocally, that confirmation of this result would give reality to the suppositions given in references [16] - [20] , which would through analysis help toward falsifiable measurements which would allow us to determine if Quantum physics and quantum gravity are, indeed part of a larger non deterministic theory, as given in [25] .

ACKNOWLEDGEMENTS

This work is supported in part by National Nature Science Foundation of China grant No. 11375279.

REFERENCES

1.   Padmanabhan, T. (2006) An Invitation to Astrophysics. World Scientific Co., Pte. Ltd., Singapore.

2.   Handley, W.J., Brechet, S.D., Lasenby, A.N. and Hobson, M.P. (2014) Kinetic Initial Conditions for Inflation. http://arxiv.org/pdf/1401.2253v2.pdf

3. Van Den Broeck, C., et al. (2015) Gravitational Wave Searches with Advanced LIGO and Advanced Virgo. http://arxiv.org/pdf/1505.04621v1.pdf

4. Cowen, R. (2015) Gravitational Waves Discovery Now Officially Dead; Combined Data from South Pole Experiment BICEP2 and Planck Probe Point to Galactic Dust as Confounding Signal. http://www.nature.com/news/gravitational-waves-discovery-now-officially-dead-1.16830

5. Cowen, R. (2014) Full-Galaxy Dust Map Muddles Search for Gravitational Waves.http://www.nature.com/news/full-galaxy-dust-map-muddles-search-for-gravitational-waves-1.15975

6. Beckwith, A. (in press) Geddankenexperiment for Degree of Flatness, or Lack Of, in Early Universe Conditions. Accepted for publication, JHEPGC.http://vixra.org/abs/1510.0108

7. Beckwith, A. (2016) "Gedanken Experiment for Fluctuation of Mass of a Graviton, Based on the Trace of a GR Stress Energy Tensor-Preplanckian Conditions Lead to Gaining of Graviton Mass, and Planckian Conditions Lead to Graviton Mass Shrinking to 10^{-62} Grams. Journal of High Energy Physics, Gravitation and Cosmology, 2, 19-24.http://vixra.org/abs/1510.0495

8. Dowker, F. (2005) Causal Sets and the Deep Structure of Space-Time. http://arxiv.org/abs/gr-qc/0508109

9. Katti, A. (2013) The Mathematical Theory of Special and General Relativity. Create Space Independent Publishing, North Charleston.

10. Haggard, H.M. and Rovelli, C. (2014) Black, Hole Fireworks: Quantum Gravity Effects Outside the Horizon Spark Black to White Hole Tunneling. http://arxiv.org/abs/1407.0989

11. Rovelli, C. and Vidotto, F. (2015) Covariant Loop Quantum Gravity: An Elementary Introduction to Quantum Gravity, and Spinfoam Theory. Cambridge University Press, Cambridge.

12. Kolb, E.W. and Turner, M.S. (1990) The Early Universe. The Advanced Book Program, Addison-Wesley Publishing Company, Redwood City.

13. Padmanabhan, T. (2010) Gravitation, Foundations and Frontiers. Cambridge University Press, Cambridge.

14. Ratra, P. and Peebles, L. (1988) Cosmological Consequences of a Rolling Homogeneous Scalar Field. Physical Review D, 37, 3406.http://dx.doi.org/10.1103/PhysRevD.37.3406

15. Goldhaber, A. and Nieto, M. (2010) Photon and Graviton Mass Limits. Reviews of Modern Physics, 82, 939-979. http://arxiv.org/abs/0809.1003http://dx.doi.org/10.1103/RevModPhys.82.939

16. Maggiorie, M. (2008) Gravitational Waves, Volume 1: Theory and Experiments. Oxford University Press, Oxford.

17. Penrose, R. (2010) Cycles of Time: An Extraordinary New View of the Universe. The Bodley Head, London.

18. Barbour, J. (2009) The Nature of Time. http://arxiv.org/pdf/0903.3489.pdf

19. Barbour, J. (2010) Shape Dynamics: An Introduction. In: Finster, F., Muller, O., Nardmann, M., Tolksdorf, J. and Zeidler, E., Eds., Quantum Field Theory and Gravity, Conceptual and Mathematical Advances in the Search for a Unified Framework, Birkhauser, Springer-Verlag, London, 257-297.

20. Beckwith, A.W. (2015) Gedankenexperiment for Refining the Unruh Metric Tensor Uncertainty Principle via Schwartzshield Geometry and Planckian Space-Time with Initial Nonzero Entropy and Applying the Riemannian- Penrose Inequality and Initial Kinetic Energy for a Lower Bound to the Graviton. http://vixra.org/abs/1509.0173

21. Corda, C. (2009) Interferometric Detection of Gravitational Waves: The Definitive Test for General Relativity. International Journal of Modern Physics D, 18, 2275-2282.http://arxiv.org/abs/0905.2502 http://dx.doi.org/10.1142/S0218271809015904

22. Capozziello, S. and Corda, C. (2006) Scalar Gravitational Waves from Scalar-Tensor Gravity: Production and Response of Interferometers. International Journal of Modern Physics D, 15, 1119-1150. http://dx.doi.org/10.1142/S0218271806008814

23. Corda, C. (2007) The Virgo-Minigrail Cross Correlation for the Detection of Scalar Gravitational Waves. Modern Physics Letters A, 22, 1727-1735. http://dx.doi.org/10.1142/S0217732307024140

24. Das, S., Mukherje, S. and Souradeep, T. (2015) Revised Cosmological Parameters after BICEP 2 and BOSS. http://arxiv.org/abs/1406.0857

25. t'Hooft, G. (1999) Quantum Gravity as a Dissipative Deterministic System.http://arxiv.org/PS_cache/gr-qc/pdf/9903/9903084v3.pdf

Chapter 2

DATA PRESERVATION AND LONG TERM ANALYSIS IN HIGH ENERGY PHYSICS

D M South

Deutsches Elektronen Synchrotron, Notkestraße 85, 22607 Hamburg, Germany

ABSTRACT

Several important and unique experimental high-energy physics programmes at a variety of facilities are coming to an end, including those at HERA, the B-factories and the Tevatron. The wealth of physics data from these experiments is the result of a significant financial and human effort, and yet until recently no coherent strategy existed for data preservation and re-use. To address this issue, an inter-experimental study group on data preservation and long-term analysis in high-energy physics was convened at the end of 2008, publishing an interim report in 2009. The membership of the study group has since expanded, including the addition of the LHC experiments, and a full status report has now been released. This report greatly expands on the ideas contained in the original publication and provides a more solid set of recommendations, not only concerning data preservation and its implementation in high-energy physics, but also the future direction and organisational model of the study group. The main messages of the status report were presented for the first time at the 2012 International Conference on Computing in High Energy and Nuclear Physics and are summarised in these proceedings.

INTRODUCTION

The last 60 years have produced a wealth of high-energy physics (HEP) results from a variety of often very different experiments. Over this period, new energy and intensity frontiers have been continuously probed, using increasingly complex accelerator installations. A growth in the size of the necessary international collaborations associated to the experiments may also be observed, as well as an increase in the diversity of the data management. The culmination of this period is not only the Large Hadron Collider at CERN, whose proton-proton collisions at a record centre of mass energy of 8 TeV will be the focus of this year's summer conferences, but also a legacy of results from many other experiments who have only recently stopped data taking and are at this time in the process of publishing their final measurements. Furthermore, it is hoped that several other proposed projects such as the Super-B factories, the ILC and the LHeC will complement the LHC physics programme in the next period.

A central question which has up to now not been properly addressed in the field of HEP is what to do with the data after the collisions have stopped and the planned analysis programme of an experiment has been completed. Until recently, there was no clear policy on this and it is likely that older HEP experiments have simply lost the data. Data preservation, including long term access, is generally not part of the planning, software design or budget of an experiment. Previous HEP data preservation initiatives have in the main not been plannedby the original collaborations, but have rather been the effort a few knowledgeable people, typically ex-collaborators, with a keen interest in preventing the data disappearing forever.

So why is it difficult to preserve HEP data, in particular seeing as a diverse range of other scientific disciplines such as astrophysics, life sciences and molecular biology have to some level integrated successful data preservation programmes into their respective fields? The scale and distribution of HEP data complicates the issue, where the question of custodianship, and who is ultimately responsible for the data - the experiments or the host computing centres - may also be unclear. Aging, unreliable hardware and highly specialised experiment-specific software contribute to the problem. Key resources, both funding and person-power expertise, tend to decrease once the data taking at a HEP experiment stops, when the focus tends to shift elsewhere. And perhaps the most important ingredient is to examine if there is a physics case to answer: can the benefits be balanced against the necessary cost and effort in developing an effective and robust data preservation programme concerning HEP data?

THE DPHEP STUDY GROUP

An international study group on data preservation and long term analysis in high-energy physics, DPHEP [1], was formed at the end of 2008 to address this issue in a systematic way. The aims of the study group include to confront the data models, clarify the concepts, set a common language, investigate the technical aspects, and compare with other fields such as astrophysics and those handling large data sets. The experiments BaBar, Belle, BES-III, CLAS, CLEO, CDF, DØ, H1, HERMES and ZEUS are represented in DPHEP; the LHC experiments ALICE, ATLAS, CMS and LHCb joined the study group in 2011. The associated computing centres at CERN (Switzerland/France), DESY

(Germany), Fermilab (USA), IHEP (China), JLAB (USA), KEK (Japan) and SLAC (USA) are all also represented in DPHEP.

Figure 1: Participants of the fifth DPHEP workshop at Fermilab, May 2011.

A series of five workshops have taken place over the last three years, and since 2009 DPHEP is officially endorsed with a mandate by the International Committee for Future Accelerators, ICFA. The initial findings of the study group were summarised in a short interim report in December 2009 [2] and a full status report was released in May 2012 [3], to coincide with the International Conference on Computing in High Energy and Nuclear Physics (CHEP) 2012 [4].

A parallel DPHEP meeting was also hosted by the conference [5]. The report contains: a tour of data preservation activities in other fields; an expanded description of the physics case; a guide to defining and establishing data preservation principles; updates from the experiments and joint projects, as well as person-power estimates for these and future projects; the proposed next steps to fully establish DPHEP in the field.

Only a brief summary of the full status report is presented in these proceedings, highlighting the areas presented in the plenary session at the CHEP 2012 conference. Additional data preservation contributions at CHEP 2012 from the BaBar, H1, HERMES and ZEUS experiments, as well as the DESY-IT division, may be also be found in the conference proceedings or on the conference webpages [4].

THE PHYSICS CASE FOR DATA PRESERVATION

When building the physics case for data preservation, four main categories have been identified: Firstly, data preservation is beneficial to the long-term completion and extension of the physics programme of an experiment. In the case of the LEP experiments a considerable tail exists in the publication rate, which continues today, and a similar trend is predicted by the HERA experiments and BaBar. Up to 10% of papers are finalised in the post-collisions period, and prolonging the availability of the data may result in a gain in scientific output of an experiment.

Secondly, cross-collaboration and combination of data from multiple experiments may provide new scientific results, with improved precision and increased sensitivity. This may occur during the active lifetime of similar experiments at one facility, such as those at LEP, HERA, or the Tevatron, but may also occur later across larger boundaries, such as combinations of Belle and BaBar or Tevatron and LHC data. The preservation of such data would facilitate the comparison of complementary physics results and allow the independent verification of experimental observations.

Thirdly, it may be useful to revisit old measurements or perform new ones with older data. Access to newly developed analysis techniques as well as the possibility to perform comparisons to state-of-the-art theoretical models may produce improved or even new physics results. Furthermore, unique data sets are available in terms of initial state particles or centre of mass energy or both, such as the PETRA e^+e^-, HERA $e^\pm p$ and Tevatron $p\bar{p}$ collision data, as well as data from a variety of fixed target experiments. More recently, the early LHC data, taken at centre of masses of 900 GeV and 2.36 TeV, as well as the low pile-up 7 TeV data taken in 2010 also provide unique opportunities.

Finally, the value of using real HEP data for scientific training, education and outreach cannot be understated. Providing a wide variety of HEP data sets for such analysis, with a corresponding wide variety of associated exercises and teaching programmes, is clearly beneficial in attracting a new generation of inquisitive minds to the field. A more detailed description of the physics case for data preservation can be found in the DPHEP status report [3], which includes specific examples of analyses using older data from among others LEP and PETRA, in addition to a description of the potential usage of the data from the experiments in the final analysis phase: the B-factories, HERA and the Tevatron.

MODELS OF DATA PRESERVATION IN HIGH-ENERGY PHYSICS

In order to develop a solid definition of models of data preservation, it is first important to ask the question: what is HEP data? That is, what needs to be preserved? The digital information, the data themselves, are clearly crucial, but at least for the pre-LHC experiments volume estimates for preservation are of the order of a few to 10 PB, which is certainly within the storage capabilities of today's HEP based computing centres. The range in data volume to be preserved is often a result not only of different sized data sets, but different types of data: from the basic level raw data, through reconstructed data, up to the analysis level ntuples.

However, if one thing may be learned from previous enterprises, it is that the conservation of tapes is not equivalent to data preservation, and that providing not only the hardware to access the data but also the software and environment to understand the data are the necessary and more challenging aspects. Therefore, in addition to the data the various software, such as simulation, reconstruction and analysis software need to be considered. If the experimental software is not available the possibility to study new observables or to incorporate new reconstruction algorithms, detector simulations or event generators is lost. Without a well defined and understood software environment the scientific potential of the data may be limited.

Just as important are the various types of documentation, covering all facets of an experiment. This includes the scientific publications in journals and online databases such as INSPIRE [6] and arXiv, published masters, diploma and Ph.D. theses, as well as a myriad of internal documentation in manuals, internal notes, slides, wikis, news-groups and so on. Detailed information about analyses may only be available in internal notes, which may not be easily available, electronically or otherwise. Many types of internal meta-data may also exist, such as the details of the detector layout and performance, hardware replacements, manuals or the documentation of meetings.

Finally, the often unique expertise and contributions of collaboration members must also be considered as another component of HEP data. Particular care is needed to ensure crucial know-how does not disappear with losses in person-power, which is liable to happen towards the end of an experiment as described in section 1.

Considering this inclusive definition of HEP data, a series of data preservation levels are established by the DPHEP study group, as summarised in table 1. The levels are organised in order of increasing benefit, which comes with increasing complexity and cost. Each level is associated with use cases, and the preservation model adopted by an experiment should reflect the level of analysis expected to be available in the future. These are the original definitions of DPHEP preservation levels from the interim report [2], which remain valid, although the interaction between the levels is now better understood: whereas the original idea was a progressively inclusive level structure, the levels are now rather seen as complementary initiatives. The four levels represent three different areas: documentation (level 1), outreach and simplified formats (level 2) and technical preservation projects (levels 3 and 4).

More generally, the implementation of a data preservation model as early as possible during the lifetime of an experiment may greatly increase the chance that the data will be available in the long term, and may also simplify the data analysis in the final years of the collaboration. Planning a transition of the collaboration structure to something more suited to a long-term organisation also makes it easier to address issues such as authorship, supervision and access.

CURRENT DATA PRESERVATION PROJECTS

Some examples of ongoing projects at the experiments and laboratories involved in the DPHEP study group are described in the following, placing them in the context of the data preservation levels described in the previous section. More details on each of the preservation levels, in addition to a full description of the various individual and joint data preservation initiatives can be found in the DPHEP status report [3].

Level 1 Projects: Documentation

Although a level 1 preservation model, to provide additional documentation, is considered the simplest, this can still require substantial activity by the experiment. The organisation of documentation requires a significant effort, with much material from pre-web days or using a range of often older web applications, and dedicated task forces have been set up by many of the experiments to perform this task. The organisation and cataloguing of non-digital documentation and the scanning or photographing of materials such as papers and talks, notes, drawings, detector schematics, blueprints, logbooks is in itself a large scale project. New "virtual archives" have been established by the experiments for the digitised material and dedicated physical storage has been secured for the newly sorted paper items.

Table 1: Data preservation levels defined by the DPHEP study group [3]

Level	Preservation Model	Use Case
1	Provide additional documentation	Publication related info search
2	Preserve the data in a simplified format	Outreach, simple training analyses
3	Preserve the analysis level software and data format	Full scientific analyses, based on the existing reconstruction
4	Preserve the simulation and reconstruction software as well as basic level data	Retain the full potential of the experimental data

Digital documentation, which encompasses online shift tools, detector configuration files, electronic logbooks and various webpages of meetings, talks, presents its own challenges. The HERA experiments have replaced old, dedicated web-servers by virtual machines (VMs), hosted by the computer centres, thus relieving the burden of updating the hardware and migrating the data in the future. The migration of old webpages to newer technologies such as wikis is also proving to be beneficial.

The use of external services for hosting collaboration material has also been examined. The internal notes from all HERA collaborations are now available on INSPIRE, so that once again the experiments no longer need to provide dedicated hardware. These INSPIRE records, which can now be linked to other records, remain password protected for now, but it is simple to make them publicly available in the future. The ingestion of other collaboration documents by INSPIRE is under discussion, including theses, preliminary results, conference talks, and proceedings. The BaBar, CDF and DØ and experiments are also working on similar projects with INSPIRE.

Level 2: Simplified Formats and Outreach

A level 2 preservation effort, to conserve the experimental data in a simplified format, is considered to be unsuitable for high level analyses, lacking the depth to allow, for example, detailed systematic studies to be performed. However, such a format is ideal for education and outreach purposes, which many experiments in the study group are also actively interested in, and may also be useful to test new theoretical models as an easy way to share complicated data.

Within DPHEP there are generic ideas, such as common formats and user interfaces. Such formats in HEP are typically based on ROOT [7], containing particle 4-vectors and simple event information, providing an input to simple analyses requiring composite-particle reconstruction or to those looking for signals in the data. Outreach initiatives are in place at BaBar, Belle and the LHC experiments and a multi-experimental project is clearly desirable, coordinated via DPHEP, and based in several locations such as CERN, FNAL and DESY, with associated tutorials linked to preserved HEP data from several sources.

Levels 3 and 4: Technical Projects

The main focus of the data preservation effort are the technical projects to preserve not only the data, but also to ensure long term access and usability. Whereas level 3 provides access to analysis level data, Monte Carlo and the analysis level software, level 4 additionally includes access to the reconstruction and simulation software. Past experiences with old HEP data indicate that new analyses and complete re-analyses are only possible if all the necessary ingredients to retrieve, reconstruct and understand the data are accounted for. Only with the full flexibility does the full potential of the data remain, equivalent to level 4 preservation. Accordingly, the majority of participating experiments in the study group plan for a level 4 preservation programme, although different approaches are employed concerning how this goal should be achieved. Typically, this can be realised in two ways: either keep the current environment alive as long as possible, or adapt and validate the environment to future changes as they happen. These two complementary approaches are taken at SLAC and DESY respectively, both employing virtualisation techniques, but in rather different ways.

The BaBar Long Term Data Access (LTDA) archival system is now installed for analysis until at least 2018. The deployment of the LTDA required a large scale investment, with over 70 new dedicated servers. Crucial to this case was that the resources for the project were taken into account in the BaBar funding model during the analysis phase. Isolated from the rest of SLAC computing, it uses virtualisation techniques to preserve an existing, stable and validated platform and provides a complete data storage and user environment in one system. A crucial part of the design is to allow frozen, older platforms to run in a secure computing environment. It is now known that a naïve virtualisation strategy is not enough, and that an old operating system (OS) cannot simply be supported forever, as the security of the system may come under threat. The LTDA safeguards against this by having a clear network separation via firewalls of the part storing the data (running a more modern OS) and the part for performing analysis (the desired older OS), as illustrated in figure 2 (top). More than 20 analyses are now using the LTDA system, where from the user's perspective it appears very similar to the classical BaBar infrastructure.

The alternative approach taken at DESY is to employ an automated validation system to facilitate future software and OS transitions. In this system, virtualisation is used for flexibility, in order to host different configurations of a software environment in one coherent set up. This is illustrated in figure 2 (bottom), where the separate inputs of experimental software, external software and OS are combined to form a VM, which is then deployed on many systems where a series of experimental tests are performed. A successfully validated environment recipe can then be installed on a future resource, such as the Grid or an IT-cluster, to provide a working environment for the experiment. An essential component of such a project is a robust definition of a complete set of experimental tests. The nature and number of the tests is dependent on desired preservation level; both H1 and HERMES aim for full level 4 preservation, ZEUS between levels 3 and 4. Providing a display of the validation results in a comprehensible way is also required, which is fulfilled using a web-based interface. The characteristics of each test determines the nature of the result, which could be simple yes/no, a set of plots, ROOT files, a text-file containing keywords or of a certain length and so on. The common OS baseline of SLD5/32-bit was achieved in 2011 by all of the HERA experiments and the ongoing validation of 64-bit systems is the next major objective towards future OS migrations. Additionally, the system has already been useful in detecting problems that were visible only in newer software.

CONCLUSION AND FUTURE WORKING DIRECTIONS

The DPHEP study group represents the first large scale effort to address data preservation in the field of high-energy physics. The initial make up of the group was driven by the coincidence of the end of data taking at several large colliders, but has grown to include others including the LHC experiments. The activity of the group over the last three years has led to an increased understanding of the relevant issues, enabling problems to be addressed, recommendations to be formulated and multi-experiment projects to begin. The recent DPHEP publication, which contains a comprehensive appraisal of data preservation in HEP, represents a significant milestone and concludes the initial period. To gain the most benefit from the work done so far, a transition from the current study group structure to a new, full time organisation is required.

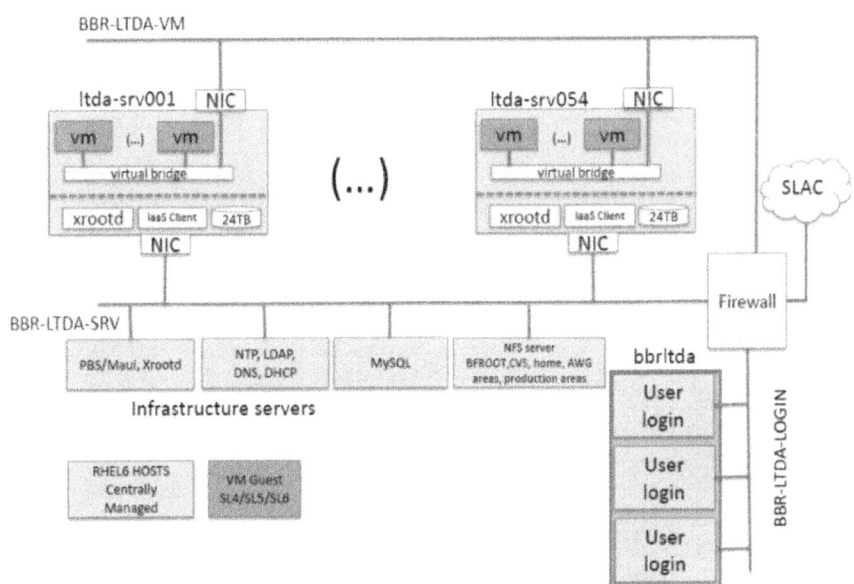

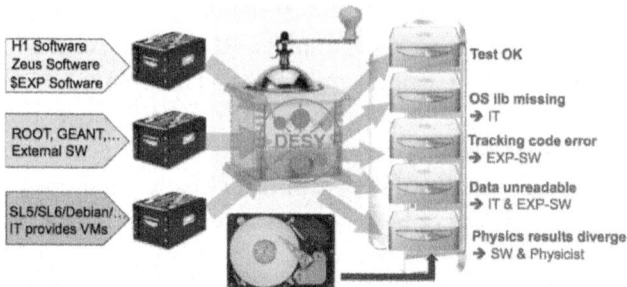

Figure 2: Top: The networking schematic of the BaBar LTDA archival system employed at SLAC. Bottom: An illustration of the generic validation framework at DESY.

The DPHEP organisation should retain the basic structure of the study group, with links to the host experiments, laboratories, funding agencies and ICFA. The main difference is the installation of a dedicated full time position, the DPHEP Project Manager, who acts as the main day-to-day operational coordinator, as illustrated in figure 3. The organisation will nevertheless continue to investigate and take action in areas of coordination, preservation standards andtechnologies, as well as expanding the experimental reach and inter-disciplinary cooperation. The next DPHEP workshop, which will focus on this last point, will take place in Autumn 2012.

Figure 3: The proposed DPHEP organisation and its associations.

REFERENCES

1. Study group on Data Preservation and Long Term Analysis in HEP, DPHEP; http://dphep.org/

2. Asner D et al (DPHEP Study Group) 2009 Preprint arXiv:0912.0255

3. Akopov Z et al (DPHEP Study Group) 2012 Preprint arXiv:1205.4667

4. 2012 International Conference on Computing in High Energy and Nuclear Physics, CHEP (New York City); http://www.chep2012.org/

5. DPHEP@CHEP2012;   https://indico.cern.ch/conferenceDisplay.py?confId=171962

6. INSPIRE is the successor to the popular SPIRES HEP database; http://inspirehep.net/

7. Brun R and Rademakers F 1997 Nucl. Instrum. Meth. A 389 (81)

Chapter 3

DATA ACQUISITION IN PARTICLE PHYSICS EXPERIMENTS

V. González[1], D. Barrientos[1, 2], J. M. Blasco[1], F. Carrió[1, 2], X. Egea[1] and E. Sanchis[1]

[1]University of Valencia, Dep. of Electronic Engineering, Spain

[2]Instituto de Física Corpuscular (IFIC), Valencia, Spain

INTRODUCTION

This chapter presents an overview of technological aspects related to data acquisition (DAQ) systems for particle physics experiments. Being a general topic as data acquisition can be, particle physics experiments pose some challenges which deserve a special description and for which special solutions may be adopted.

Generally speaking, most of the particle physics experiments incur in the use of different types of sensors which information has to be gathered and processed to extract the biggest semantic contents about the process being analyzed. This fact implies the need for, not only a hardware coordination (timing, data width, speeds, etc.) between different sub-acquisition systems for the different types of sensors, but also an information processing strategy to gain more significance from the analysis of the data from the whole system that the one got from a single type of sensor. Also, from the point of view of hardware resources, each type of sensor is replicated several times (even millions) to achieve some spatial coverage. This fact directly drives to the extensive use of integrated devices when needed to improve cost and space utilization.

This chapter, thus, will cover the specific technologies used in the different stages in which a general DAQ system in particle physics experiments can be divided.

The rest of the chapter is organized following the natural flow of data from the sensor to the final processing. First, we will describe the most general abstraction of DAQ systems pointing out the general architecture used in particle physics experiment. Second, different common types of transducers used will be described with their main characteristics. A review of the different hardware architectures for the front-end system will follow. Then, we will get into several common data transmission paradigms including modern standard buses and optical fibers. Finally, a review of present hardware processing solutions will be done.

DATA ACQUISITION ARCHITECTURES

Data acquisition pursues the reading of the information from one or many sensors for its real-time use or storage and further off-line analysis. Strictly speaking we may establish four different activities in a sensor processing system: acquisition, processing, integration and analysis. However, most of the time we refer to DAQ system as the whole of these activities.

It is worth to say that not every DAQ system includes these four activities, depending on their complexity and application. For example, in single sensor systems neither integration nor processing could be necessary. On the other hand, in systems with replicated sensors, processing could be minimal, but the integration is crucial. If the system is based on different types of sensors, processing is necessary to make the readings of the various sensors compatible and the integration is needed to obtain comprehensive information of the environment. However, the majority of the DAQ systems will include the four activities: the physical variable is sensed in the acquisition activity; the data collected is processed properly (for example, performing scaling or formatting) before being transported to the integration activity; the output of the integration is a more meaningful information on which the analysis activity can base its tasks (storage, action on a mechanism, etc.).

Architectures of Sensor Systems

As we mentioned before, the DAQ system consists of four activities: acquisition, processing, integration and analysis. Depending on the characteristics of the process under study we will have to choose how to organize them, as we shall see now, to adapt it to our needs [1].

- Collection of sensors

A collection of sensors is a set of sensors arranged in a certain way. They can be in series, parallel or a mixed combination of these two basic arrangements.

The choice of the particular configuration will depend on the application. The integration of information is carried out progressively through the different sensors to get a final result.

- Hierarchical system

In a centralized system, data from the sensors are transmitted to a central processor to be combined. If the volume of data is large, this organization may require considerable bandwidth. For these cases, the DAQ system can be arranged as a hierarchy of subsystems.

Consider the example of the figure 1a. Data D_1 and D_2 are combined in the stage of integration in the feature F_{12}. Similarly, D_3 and D_4, D_5 and D_6 and D_7 and D_8 constitute characteristics F_{34}, F_{56} and F_{78} respectively. Features F_{12} and F_{34} are then combined in the stage of feature integration in a local decision *Dec 1-4*, and F_{56} and F_{78} generate *Dec 5-8*. Finally, local decisions are combined in the stage of decision integration in a global decision *Dec 1-8*.

The interesting aspect of this organization is that an increase in the "size" of the problem does not translate in a similar increase in the organization of the DAQ, i.e., the system does not grow linearly with the problem. This is true provided that the data and feature fusion stages reduce the volume of information.

- Multisensor integration

Figure 1b shows the integration of various sensors s_1, s_2, s_3 and s_4 (not all of the same type). We assume that s_1 and s_2 are of the same type, s_3 of a second type and s_4 of a third type. In this case, the integration of the information has to be done to ensure that data from the sensors are compatible.

The processing of the four readings must be carried out sequentially in three phases: $t_1 = F_1\ (s_1,\ s_2)$, $t_2 = F_2\ (t_1,\ s_3)$ and $t_3 = F_3\ (t_2,\ s_4)$. The final output is $t_3 = F_3\ (F_2\ (F_1\ (s_1,\ s_2),\ s_3),\ s_4)$. We must clarify that the function F_1 uses data from the two sensors competitively while F_2 and F_3 use complementary data.

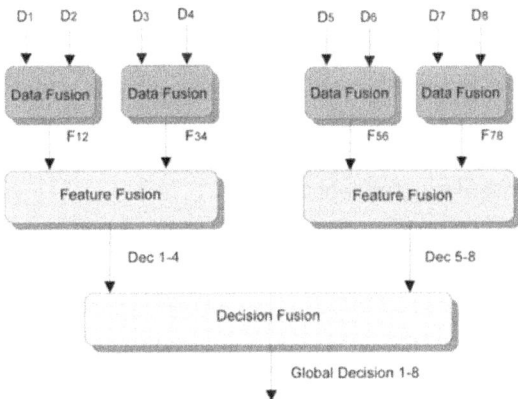

a

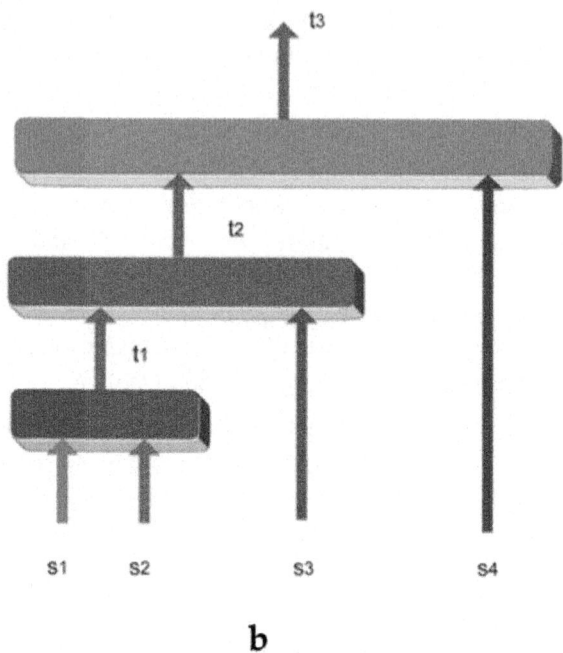

b

Figure 1: a) Hierarchical sensor integration. b) Multisensor integration.

Distributed Processing of Sensors

In the following we will focus on DAQ systems where the four activities described before take place in a distributed way, known as distributed processing systems [2]. This case is of special interest as many of the present DAQ systems for particle physics follow it.

To better understand the different aspects of the processing of sensors let us consider a distributed processing system of sensors with the main objective of detecting targets present in the surveyed space. This example may apply to particle physics experiments but also to other fields like distributed control systems, sensory systems for robots, etc.

Let us assume that there is a finite number of resources (sensors and processors) in the distributed system. Consider a system in which there are N sensors (S_1 to S_N) and P processors (EP_1 EP_P). N sensors, for example, can track objects in observation space and we assume that they are all of the same type, that is, they conform a system based on physically replicated sensors. Let us suppose they have been organized in P sensors groups or *clusters*, for example 3, of N/P sensors each. In our example, there are three groups, each with three sensors and a processor to control them.

The main task, T, is to detect and possibly follow targets across the surveyed space. Consider two possibilities:

- The space of observation is too broad and therefore cannot be efficiently covered by any of the clusters of sensors.
- The space of observation can be covered by all the groups of sensors, but the system requires a response in real time for the follow-up of the target.

In the first case, part of the space can be assigned to each group of sensors. Collectively, they will cover all the surveyed space. In the second case, we can assign to each cluster the task of following some specific number of targets; ideally, each group should be following a single target.

In our example, the distributed processing system breaks down the main task T in P subtasks; this operation is known as *task decomposition*. The objective of each subtask T_i is to detect and follow the ith-target in the observation space. Each task is assigned with a processing element, EP_i, that controls the three sensors of the group.

Each group of replicated sensors has a local processor. The processor is responsible for local processing and control; it can control the sensors assigned to it and obtain the values from them. Ideally, the sensors of the clusters should always obtain the same value, but in practice they give different values following some statistical distribution.

Suppose that each group can see only part of the space, but the targets can move anywhere within this space. In this case, the system would require a communication between the local processors to share the information about the object and to know when it moves from one area to another.

Finally, the integrator is responsible for combining data from sensors and/or their abstractions. It should be noted that we started with nine sensors. There are three groups of three sensors each. The three sensors on each group provide redundant information. The processing of each group combines the redundant information to obtain a solution of a sub-problem - what object is the one the group is observing?

In this way, the integrator gets three sets of data, each coming from a group of sensors. With these data, the observer determines that there are three objects in the space of observation. The distance from the integrator to the sensor is not, in general, negligible, so the results of the local processing must be transmitted in some way.

In our example, the result obtained by the integrator is a map of the objects present in the whole surveyed space.

The DAQ system is assumed to have a knowledge base that can analyze and interpret the data and take the appropriate action depending on the result obtained. In our case, the system interprets that there are three objects that occupy the space of observation; the reaction of the system will depend on the knowledge base.

Distributed Sensor Networks

The use of different, intelligent and distributed sensors space and geographically has grown constantly in applications such as robotics, particle physics experiments, medical images, tracking radar, air navigation and control systems of activities on production lines, to name a few. These systems, and other similar, are called *distributed sensors networks* or *DSN* [1]. Otherwise, we could define a network of distributed sensors as a set of intelligent sensors distributed spatially and designed to obtain data of the environment that surrounds them, abstracting the relevant information and infer from it the observed object, deriving from all this, an action appropriate according to the scenario.

Data Acquisition Systems in Particle Physics Experiments

The distributed sensor network (DSN) paradigm fits what we generally implement as DAQ systems in particle physics experiments. Because of the need of a spatial coverage or an identification scheme based on the detection of different types of particles, the DAQ system will include several sensors of the same type or different types of multiple replicated sensors. Hardware architectures to read out all them are implemented in a distributed and possibly hierarchical way due to high data volume, high data rate or geographical sensor distribution. Comparison of hierarchical DSN versus other type of solution may be found in [3, 4].

RADIATION DETECTION. TRANSDUCERS

Radiation detection involves the conversion of the impinging energy in form of radiation into an electrical parameter which can be processed. In order to achieve this, transducers are the responsible for transforming the radiation energy into an electrical signal. The type of detector has to be specific for each radiation and its energy interval. In general, several factors must be taken into consideration as the sensitivity, the response of the detector in energy resolution, response time and efficiency of the detector.

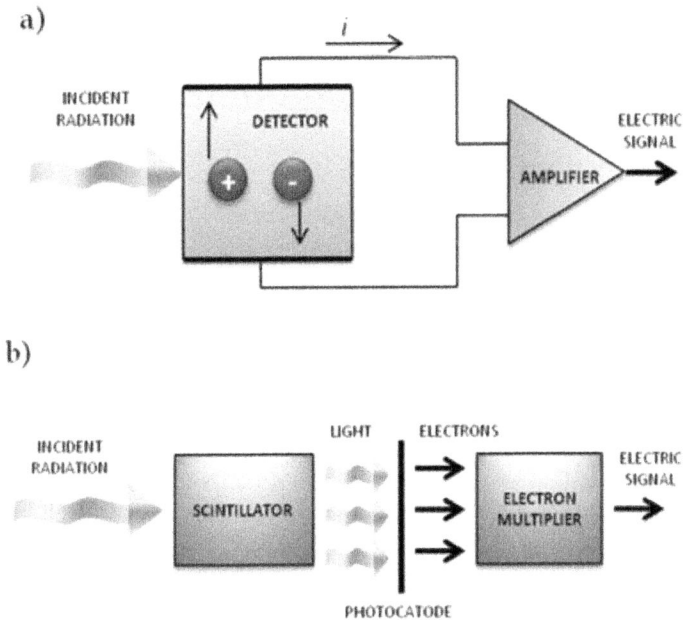

Figure 2: Radiation detection. (a) direct mode, (b) indirect mode.

Energy conversion can be carried out whether in a *direct* mode, if the signal is directly detected through the ionization of a material (figure 2a), or in an *indirect* mode, when it performs different energy conversions before obtaining the electrical signal (light production plus electrical conversion,figure 2b). The following sub-sections describe the most commonly used devices for both medical as nuclear physics applications.

Direct Detection

Direct detection with ionization chambers is a common practice. They are built with two electrodes to which a certain electrical potential is applied. The space between electrodes is occupied by a gas and it responds to the ionization produced by radiation as it passes through the gas. Ionizing radiation dissipates part or all of its energy by generating electron-ion pairs which are put in motion by the influence of an electrical field and consequently, by producing an electrical current.

Other possibility that provides good results in radiation detection is a semiconductor detector. They are solid-state devices which operate essentially like ionization chambers but in this case, the charge carriers are electron-hole pairs. Nowadays, most efficient detectors are made of Silicon (Si) or

Germanium (Ge). The main advantage is their high energy resolution; besides, they provide linear responses in a wide energy range, fast pulse rise time, several geometric shapes (although the size is limited) and insensitivity to magnetic fields [5].

Indirect Detection

Scintillators

Scintillators are materials which exhibit luminescence when ionizing radiation passes through them. Material absorbs part of the incident energy and reemits it as light, typically in the visible spectrum. Sir William Crooks discovered this property presented by some materials in 1903 when bombarding ZnS with alpha particles.

Organic scintillators belong to the class of the aromatic compounds like benzene or anthracene. They are made by combining a substance in higher concentration, solvent, and one or more compounds at lower concentrations, solutes, which are generally responsible for the scintillation. They are mainly used for the detection of beta particles (fast electrons with linear response from ~125 keV), alpha particles and protons (not linear response and lower efficiency for the same energies) and also for the detection of fast neutrons. They can be found in different states such crystals, liquid solutions, scintillating plastics with almost every shape and size and in gaseous state.

On the other hand, inorganic scintillators are crystals of the alkali metals such as NaI(Tl), Cs(Tl), LiI(Eu) and $CaF_2(Eu)$. The element in brackets is the activator responsible of the scintillation with a small concentration in the crystal. Inorganic scintillators have in general high Z and for this reason they are mainly used for gamma particle detection, presenting a linear response up to 400 keV. Regarding to its behavior towards charged particle detection, they exhibit linear responses with the energy of the protons from 1 MeV and for alpha particles from 15 MeV. However, they are not commonly used to detect charged particles [5,6,7].

As it is shown in figure 2b, the scintillator produces a light signal when it is crossed by the radiation to be detected. It is coupled to a photodetector that will be responsible of transforming the light signal into an electrical signal.

Optoelectronic Technology for Radiation Detection

Light detection is achieved with the generation of electron-hole pairs in the photosensor in response to an incident light. When the incident photons have energy enough to produce photoelectric effect, the electrons of the valence band jump to the conduction band where the free charges can move along the material under the influence of an external electric field. Thus, the holes left in the valence band due to prior removal and displacement of electrons, contribute to the electrical conduction and in this way, photocurrent is generated from the light signal.

Photodetectors. Features

One of the main characteristics of a photodetector is its spectral response. The level of electric current produced by the incidence of light varies according to the wavelength. The relationship between them is given by the spectral response, expressed in form of photosensitivity S (A/W) or quantum efficiency QE (%). Another important feature is the Signal to Noise (SNR) Ratio. It is a measure that compares the level of a desired signal to the level of the background noise.

The sensibility of the photodetector depends on certain factors such as the active area of the detector and its noise. The active area usually depends on the construction material of the detector; about the noise level, it is expected that the level of the signal exceeds the noise associated to the detector and its electronics, taking into account the desired SNR. One important component of the noise in the photodetector is the dark current [8, 9]; this current is due to the current flow existing in the photodetector even when they are in a dark environment, both in the photoconductive mode as in the photovoltaic mode. This current is known as dark current with intensities from nA to pA depending on the quality of the sensor.

The light coming from the scintillator is generally of low intensity and because of that, some photodetectors make avalanche processes to multiply the electrons for obtaining a detectable electric signal. Other parameters that determine the quality of the photodetector are the reverse voltage, the time response and its response against temperature fluctuations.

Commercial Photodetectors

Photomultiplier Tubes (PMT) have been the photodetectors longer employed for a wide number of applications, mainly due to their good features and benefit results. They are used in applications that require measuring low-level light signals, for example the light from a scintillator, converting few hundred photons into an electrical signal without adding a large quantity of noise. A photomultiplier is a vacuum tube that converts photons in electrons by photoelectric effect. It consists of a cathode made of photosensitive material, an electron collecting system, some dynodes for multiplying the electrons and finally an anode which outputs the electrical signal, all encapsulated in a crystal tube. The research carried out in this type of detectors and the evolutionary trend is mainly focused on the improvement of the QE, achieved with the development of the photomultipliers with bialkali photocathode or GaAsP, but it has also been focused on obtaining better time response. In relation to the building material, four of them are commonly used depending on the detection requirements and the wavelength of the light, (Si, Ge, InGaAs and PbS, figure 3).

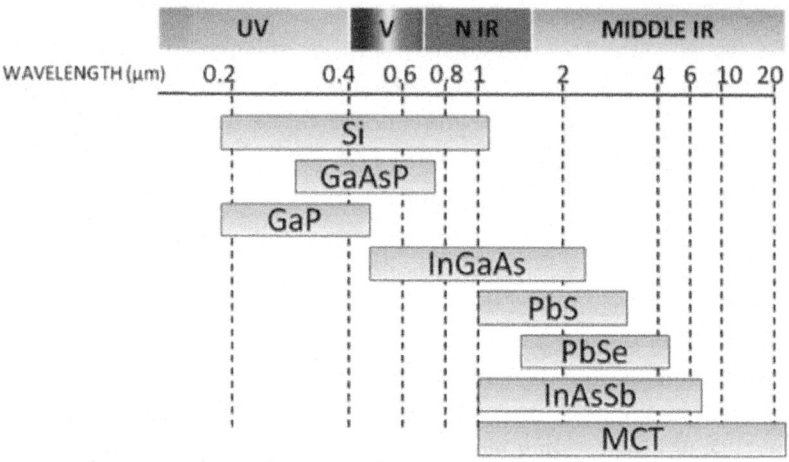

Figure 3: Sensitive material according to wavelength.

In such applications where the level of light is high enough, photodiodes use to be the detectors employed due to their lower price but also to their remarkable properties and response. It is a semiconductor with a PN junction sensitive to the infrared and visible light. If the light energy is greater than the band gap energy, the electrons are pulled up into the conduction band, leaving holes in their place in the valence band. If a reverse bias is applied, then there is an electrical current. Thus, P layer in the surface with the N layer both act as a photoelectric converter.

Other important photodetectors such as Avalanche Photodiodes (APDs) have been developed in the last few years [10]. Compared to photodiodes, APDs can detect lower levels of light and they are employed in applications where high sensitivity is required. Although the principle of operation, materials and construction are similar to the photodiodes, there are considerable differences. It has an absorption area A and a multiplication area M which implies an internal gain mechanism that works by applying a reverse voltage. When a photon strikes the APD, electron-hole pairs are created and in the gain area, the acceleration of the electrons is produced; thus, the avalanche process starts with a chain reaction due to successive ionizations. Finally, the reaction is controlled in a depletion area. The output result is, after the incidence of a photon, not only the generation of one or few electrons but a large number of them. In this way, a high level of electric current is obtained from a low level of incident light with gain values around 10^8.

Silicon Photomultipliers (Si PMT) are promising detectors due to their characteristic features and probably its utilization in many applications will increase during the following years. It is a photon counting device consisting on multiple APD pixels forming an array and that operates in Geiger mode. The addition of the output of each APD results on the output signal of the device, allowing the counting of individual events (photons). One advantage is the low reverse voltage needed for its operation, lower than the one used with PMTs and APDs. When the reverse voltage applied exceeds the breaking reverse voltage, the internal electric field is high enough to produce high gains of the order of 10^6 [11, 12].

Finally, a CCD camera (charged-coupled device) is an integrated circuit for digital imaging where the pixels are formed with p-doped MOS capacitors. Its principle of operation is based on the photoelectric effect and its sensitivity depends on the QE of the detector. At the end of the exposition, the capacitors transmit their charge which is proportional to the amount of light and the detector is read line by line (although there are different configurations). They offer higher performances regarding QE and noise levels; however their disadvantages are the big size and high price. Figure 4 shows the general features of the different photodetectors.

Photo detector	λ (nm)	Gain	Rise Time (ns)	Active Area (mm²)	V (V)	QE (%)
PD	200-2000	-	2-20	$0.5-2 \cdot 10^3$	5-15	70-95
PMT	115-1100	10^3-10^7	0.7-15		300-3000	1-80
APD	300-1700	10-10^8	1-4	0.2-5	150-1500	50-85
Si PMT	400-550	10^5-10^6	~1	0.5-3	50-100	25-65
MCP	100-650	10^3-10^7	0.15-0.8	$5-15 \cdot 10^{-3}$ /channel	500-3500	~1
HPD	115-850	10^3-10^5	0.4-10	2-10	~10^4	25-45
CCD	200-1200	-	-	$5-15 \cdot 10^{-3}$ /pixel	5-15	70

Figure 4: General features of different photodetectors.

FRONT-END ELECTRONICS

When talking about front-end electronics in nuclear or particle physics applications, we usually refer to the closest electronics to the detector, involving processes from amplification, pulse-shape conformation to the analog-to-digital conversion. The back-end electronics are left apart further away from the detector for processing tasks.

In this section, we will introduce the common circuits used in the front-end electronics, such as preamplifiers, shapers, discriminators, ADCs, coincidence units and TDCs.

Unipolar and Bipolar Signals

In nuclear and particle physics, usually the signals obtained are pulse signals. Depending on the detector used, different parameters such as the rise or fall time, as well as the amplitude are different.Figure 5 (left) shows a typical pulse signal with all its important parameters. Mostly related to the rise time, it is important to remark the bandwidth of pulse signals, related to the fastest component of the pulse, usually the rise time. A typical criteria to choose the signal bandwidth based on the temporal parameters is to choose a bandwidth such that BW=0.35/t_r, where t_r is the signal rise time [13].

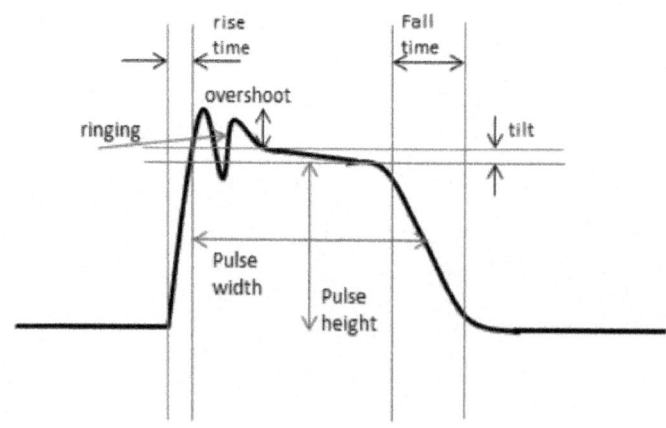

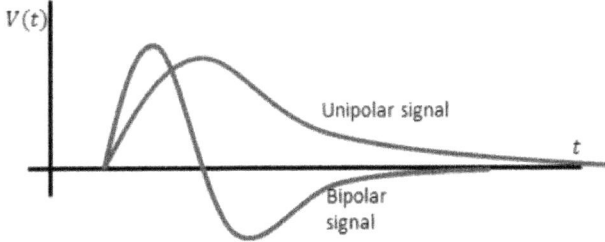

Figure 5: Left. Example of a typical pulse signal. Right. Example of a bipolar and unipolar signal.

Unipolar or bipolar signals show better or worse performance depending on our needs, each of them with different advantages and drawbacks. By definition, a unipolar signal is an only positive or only negative signal while a bipolar signal involves both signs, more interesting for high-counting rate systems, while the unipolar is left for systems with lower counting rates, with a better SNR. On figure 5 (right), an example of a unipolar and bipolar signal can be seen. For more information about this, consult the references [6, 7].

Circuits

Preamplifiers

Often in particle physics nuclear and particle physics experiments, the signal obtained at the output of the amplifier is an electrical pulse whose amplitude is proportional to the charge produced by the incident radiation energy. It is quite impractical to provide directly the signal without a proper amplification, and for this reason, preamplifiers are the first stage seen by the pulse signal, usually placed the closest to the detector for noise minimization since noise at this stage is very critical. Two different types of preamplifiers are commonly used depending on the sensing magnitude: Voltage-sensitive amplifiers and charge-sensitive amplifiers.

Voltage-Sensitive Amplifiers

They are the most conventional type of amplifiers, and they provide an output pulse proportional to the input pulse, which is proportional to the collected charge as well. If the equivalent capacitance of the detector and electronics is constant, this configuration can be used. On the other hand, in some applications, as for example semiconductor detectors, the detector capacitance changes with temperature, so this configuration is not anymore useful. Hence, it is preferred to use the configuration called charge-sensitive preamplifiers. The basic schematic of the voltage-sensitive amplifier is shown on the figure 6 (left).

Charge-Sensitive Amplifiers

Semiconductors, such as germanium or silicon detectors are capacitive detectors itself with very high impedance. The capacitance, C_i, for these detectors fluctuates making the voltage-sensitive amplifier inoperable. The idea of this circuit is to integrate the charge using the feedback capacitor C_f. The advantage of this configuration is the independence of the amplitude with the input capacitance if the condition $A \ll (C_i + C_f) C_f$ is satisfied. A picture of the charge-sensitive amplifier is shown on the figure 6 (right) [14].

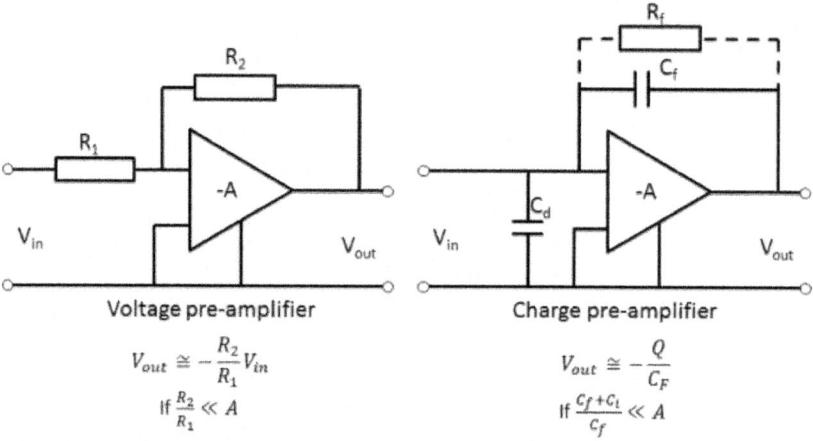

Figure 6: Voltage and charge preamplifiers schematic.

The feedback resistor R_f is used to discharge the capacitor leading the signal to the baseline level with an exponential tail around 40-50 μs. This discharge is usually done with a high R in order to provide a slow pulse tail, minimizing the noise introduced, but a tail too slow can lead to pile-up effects. Another approach to get rid of the pile-up effect is the optical feedback charge amplifier [6, 7].

Amplifiers and Shapers

After the pre-amplification process is carried out, it might be useful to provide a certain shape in order to simplify the measurements of certain magnitudes, preserving the interest magnitude intact. Pulse stretching and spreading techniques can be used for pile-up cancellation, timing measurements, pulse-height measurements and preparation for sampling. Other reasons to use pulse shaper is its SNR optimization, where a certain shape provides the optimal SNR ratio. Most of the shaper circuits are based on differentiator (CR) and integrator (RC) circuits. The circuit schematic and time response are shown in the figure 7 (left). For further information about these circuits, consult the references [7, 15].

Shaper Networks

Three different pulse shapers will be introduced into this sub-section, although there exist many more. CR-RC network, CR-RC network with pole-zero cancellation and the double differentiating CR-RC-CR circuit are introduced here. CR-RC circuits are implemented as a differentiator followed by an integrator (figure 7a). The differentiated pulse allows the signal to return to the baseline level but it does show neither an attractive pulse nor allows easy sampling of the maximum point when extracting the energy with pulse height analysis (PSA). The integrator stage improves the SNR ratio and smoothens the waveform.

The choice of the time constant often is a compromise between pile-up reduction and the ballistic deficit, which occurs when the shaper produces an amplitude drop. This can be solved by choosing a high value of ompared to the rise time, or the charge collecting time from the detector.

When considering a real pulse instead of a step ideal response, CR-RC circuits produce undershoot (figure 7b), which leads to a wrong amplitude level. This can be solved by adding a resistor to cancel the pole of the exponential tail, cancelling the undershoot. If the system counting rate is low, this strategy is useful, but when this counting rate increases, the pulses start to pile-up onto each other creating baseline fluctuations and amplitude distortion.

A solution for this problem is the double differentiating network CR-RC-CR (figure 7c), in which a bipolar pulse is obtained from the input pulse. The main difference resides on the fact that the bipolar pulse does not leave any residual charge, making it very suitable for systems with high-counting rates, but still for systems with low-counting rates, it is preferred to use the unipolar pulses, since its SNR ratio is fairly better.

Further shaping methods such as semi-gaussian shapers, active pulse shapers, triangular and trapezoidal shaping, as well as shapers using delay lines can be consulted on the references [6, 7, 15].

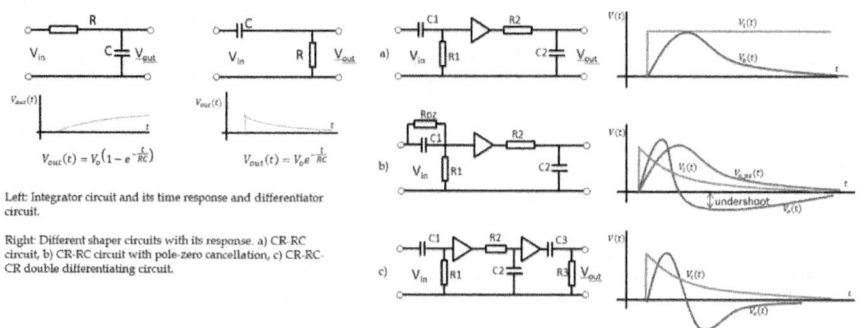

Figure 7: Left. Integrator and differentiator circuits. Right. a) CR-RC circuit, b) CR-RC circuit with pole compensation, c) CR-RC-CR double differentiating circuit.

Discrimination Techniques

Discriminator circuits are systems that are activated only if the amplitude of the input signal crosses a certain threshold. Discriminators are used to find the events and to use them as trigger signals, commonly for time measurement. Besides, it blocks the noise coming from previous devices, such as the detector and other electronics stages.

The simplest method for pulse discrimination is the leading edge triggering. It provides a logic signal if the pulse amplitude is higher than a threshold. The logic signal is originated at the moment when the signal crosses the threshold but with the problem called the time-walk effect, which describes the dependence of the pulse discrimination with the signal rise time. This effect can be seen, on the figure 8 left.

Another undesirable effect for pulse discrimination is the time jitter effect (figure 8 right). This effect is caused by statistical fluctuations at the detector and electronics level, and as a difference from the time walks effect. The time jitter is shown as a timing uncertainty when the signal amplitude is constant. This effect comes from the noise introduced in the components, and also the detector sources, as for example the transit time from the electrons in a photomultiplier or the fluctuation of photons produced in a scintillator.

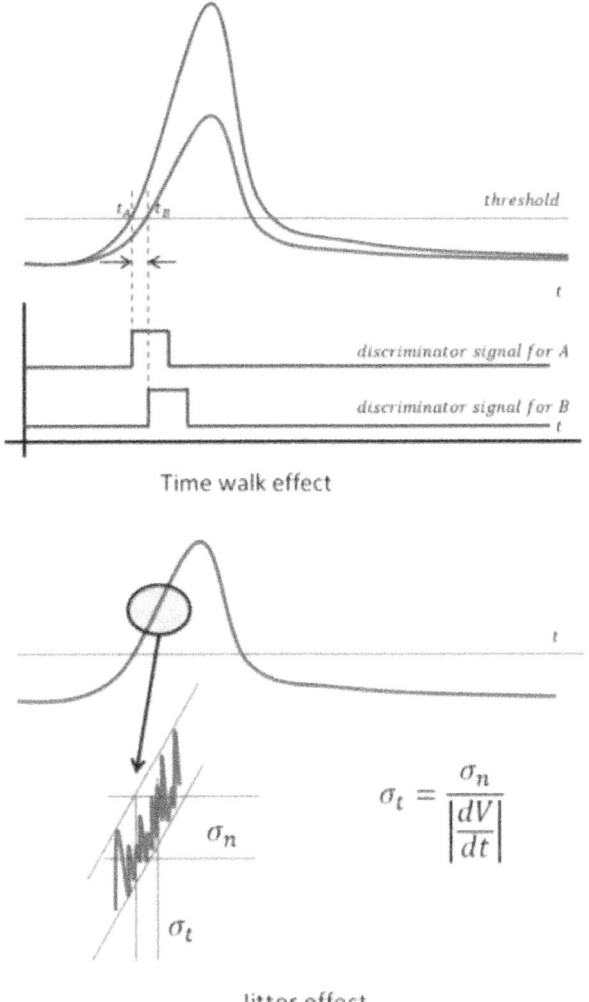

$$\sigma_t = \frac{\sigma_n}{\left|\dfrac{dV}{dt}\right|}$$

Figure 8: Time walk and jitter effects.

Other methods to avoid or reduce these effects in discrimination systems are zero-crossover timing or constant fraction discrimination methods. Zero-crossover timing method is based on the double-differentiation of the pulse shape. This method, although improves the time resolution and makes independent the crossing point from the amplitude, the shape and rise time still influence the time resolution, making it unsuitable for applications where these fluctuations are very large. The constant fraction discriminators establish the threshold as a fraction of its maximum level. The most common way to implement it is based on the comparison between a fraction of the signal with a slightly delayed version, where the zero-crossing point of its difference makes the pulse independent of the amplitude with lower jitter [7].

A/D Conversion (Analog to Digital Conversion)

More sophisticated algorithms may be implemented digitally inside the logic devices (FPGA or GPUs). Nevertheless, before performing those algorithms, an analog to digital conversion is required, introducing inevitably a source of error due to the sampling and quantization processes. Two of the most common techniques used in nuclear or particle physics research proposals are the Wilkinson method and the FADC (Flash ADC) if a very high sampling rate is required, although other conversion methods such as successive approximation and sub-ranging ADC are used as well.

The main difference between the Wilkinson method and FADC sampling is that Wilkinson method takes one sample per event based on the time measured when a capacitor is discharged, where the time counted is proportional to the pulse charge. On the other hand, FADC takes several samples per event, where the digitized value is taken when comparing the input voltage with a set of resistors forming a voltage divider across all the possible digitized values. Although FADC technique leads to the fastest architecture, as far as the number of bits required is higher, the amount of comparators increases exponentially [6, 16, 17].

The analog-to-digital conversion performance can be tested by measuring certain parameters, such as the differential and integral nonlinearities (DNL and INL), which cause missing codes, noise and distortion, as well as the effective number of bits (ENOB), which quantifies the resolution loss when the distortion and nonlinearities come in. Further information about ADC parameters con be found in [17] and its measuring method in [17, 18].

Coincidence and Anti-Coincidence Units

These circuits are used to know whether an event has been detected in several detectors at the same time or to detect events only occurring at only one detector. This is especially useful in detector arrays in order to discard fake events. The way to implement it, is based on simple logical operations between the signals from the discriminators [6, 7].

TDC (Time-To-Digital Converter)

In most of the applications in nuclear and particle physics, the measurement of time intervals will be a primary task. Basically the way to measure time intervals, is based on a start and stop signals, usually given by discriminator circuits. Then, a value proportional to the time interval between the start and stop signals is digitized. Different architectures lead to different performance, but the most notorious for TDC is the time resolution, defined as the minimum time the TDC is capable to measure. Among different architectures, we can mention the TAC (Time-to-amplitude converter), the direct time-to-digital converter, and for higher resolutions, the differential TDC and the Vernier counter [15].

DATA BUS SYSTEMS FOR BACK-END ELECTRONICS

In this section we present the most popular standards used today to build DAQ systems in particle physics experiments. All the systems presented are modular systems with each module carrying out a specific function. This technique allows the reuse of the modules in other systems and makes the DAQ system scalable.

Most of the features of these modular systems as mechanics, data buses characteristics or data protocols are defined in standards. Many DAQ system manufacturers develop their own products according to these standards. The use of standards implies many benefits as the use of third party products and support [19].

NIM Standard

NIM stands for Nuclear Instrumentation Module and was established in 1964. NIM standard does not include any kind of bus for data transfer communications since NIM crates only provide power to the NIM modules.

The advantage of the NIM standard is that modules are interchangeable and work as a standalone system allowing the set up of DAQ systems in a simple way, where a module can be replaced without affecting the integrity of the rest of the system [20]. These advantages make the NIM standard very popular in nuclear and particle experiments, and it is still used for small experiments. However, NIM has disadvantages since lacks of a digital bus, not allowing a computer based control or data communication between modules.

Crate and Modules

Standard NIM crates have 12 slots for modules and include a power supply that provides AC and DC power to the modules. The power supply is distributed via the backplane to the NIM modules that comprise many different functions like discriminators, counters, coincidences, amplifiers or level converters, for example.

Figures 9a and 9b show a standard NIM crate where we can see the NIM connector in the bottom, and a standard NIM module.

a

b

Figure 9: a) Standard NIM crate, b) Standard NIM module.

VME Standard

The Versa Module Europa (VME) is a standard introduced by Mostek, Motorola, Philips and Thompson in 1981. It offers a backplane that provides fast data transfer allowing an increase of the amount of transferred data and, therefore, an increase of the channel count coming from the front-end electronics. This fact makes VME standard the widest standard used in physics experiments.

Crate and Modules

VME crates contain a maximum of 21 slots where the first position is reserved for a controller module; the other 20 slots are available for modules that can perform other functions.

There are different types of VME modules, each having a different size and a different number of 96-pin connectors that define the number of bits designed for address and data buses.

VMEBUS

VME systems use a parallel and asynchronous bus (VMEBus) with a unique arbiter and multiple masters. VMEBus also implements a handshaking protocol with multiprocessing and interrupt capabilities. VMEbus is composed by four different sub-buses: Data Transfer Bus, Arbitration Bus, Priority Interrupt Bus and Utility Bus [20].

While VMEBus achieves a maximum data transfer of 40 MBps, some extensions of the VME standard as VME64, VME64x and VME320 standards have enhanced its capabilities increasing the number of bits for address and data and implementing specific protocols for data communication [21]. In this way, VME64 systems achieve data transfers up to 80 MBps, while VME64x support data transfers up to 160 MBps and VME320 between 320 MBps and 500 MBps.

Also, VXS standard is an ANSI/VITA standard approved in 2006. VXS standard maintains backward compatibility with VME systems combining the parallel VMEbus with switched serial fabrics. VXS systems achieve a maximum data transfer between modules of 3050 MBps [22].

PCI Standard

PCI stands for Peripheral Component Interface and was introduced by Intel Corporation in 1991. The PCI bus is the most popular method used today for connecting peripheral boards to a PC providing a high performance bus using 32 bit or 64 bit bus with multiplexed address and data lines. PC-based DAQ systems can be easily built using PCI systems as PCI cards are directly connected to a PC.

PCI Cards

The last PCI standard specifies three basic form factors for PCI cards: long, short and low profile [23]. PCI cards are keyed to distinguish between 5V or

3.3V signaling cards and they use different pin count connectors according to the data and address bus widths.

PCI Local Bus

PCI devices are connected to the PC via a parallel bus called PCI Local Bus. Typical PCI Local Bus implementations support up to four PCI boards that share the address bus, data bus and most of the protocol lines, but also having dedicated lines for arbitration. PCI local bus width and clock speed determines the maximum data transfer speed. Table 1 shows a summary of the achievable data transfer speeds in PCI and the extension version of the PCI, called PCI-X [24].

A disadvantage of the PCI standard is the use of a parallel bus for data and address lines. The skew between lines and the fact that only one master/slave pair can communicate at any time and the handshaking protocol limits the maximum achievable data transfer in PCI [19].

Further, in 1995, PICMG introduced the compact PCI (cPCI) standard as a very high performance bus based on the PCI bus using Eurocard format boards. But cPCI is not widely used in particle physics experiments due to some additional disadvantages, such as, small size cards, limited power consumption and limited number of slots [25].

Table 1: Data transfer speeds for PCI and PCI-X standards

Standard version	Bus width	Clock speed (MHz)	Data rate (Mbps)
PCI 1.0	32	33	133
PCI 2.1	32	66	266
	64	66	532
PCI-X 1.0	64	133	1064
PCI-X 2.0	64	266	2128
	64	533	4256

PCI Express

PCIe stands for Peripherial Compatible Interface Express. PCIe standard was introduced in 2002 to overcome the space and speed limitations of the conventional PCI bus by increasing the bandwidth while decreasing the pin count. This standard not only defines the electrical characteristics of a point to point serial link communication, but also a protocol for the physical layer, data link layer and transaction layer. Moreover PCIe standard includes advanced features such as active power management, quality of service, hot plug and hot swap support, data integrity, error handling and true isochronous capabilities [26, 27]. As PCI systems, PCIe systems allow the implementation of DAQ systems based on PC.

PCIE Cards

PCIe uses four different connector versions: x1, x4, x8 and x16, where the number refers to the number of available bi-directional data path and correspond to 32, 64, 98 and 164 pin connectors respectively. There are two possible form factors for PCIe cards: Long and Short.

PCIE Bus

The PCIe serial bus transmits with a data rate of 2.5 Gbps using LVDS logic standard. But really, the effective data rate is reduced to 80% of the original data rate due to the use of the 8b10b codification [27]. A summary of the achieved data rates per direction using PCIe is shown in table 2:

Table 2: Data transfer speeds for the PCIe standard

Bus	Data rate (MBps)
PCI-Express x1	250
PCI-Express x4	1000
PCI-Express x8	2000
PCI-Express x16	4000

ATCA Standard

ATCA stands for Advanced Telecommunications Computing Architecture and was introduced by PICMG in 2002 in the PICMG 3.0 specification. PICMG 3.0 and 3.x specifications define a modular open architecture including mechanical features, components, power distribution, backplane and communications protocols. This specification was created for telecommunication purposes where high speed, high availability and reliability are extremely needed. ATCA systems can deploy a service availability of 99.999% in time [28].

ATCA Shelf

The shelf can allocate a different number of ATCA boards (blades) and it also allocates the shelf manager that is responsible for the power and thermal control issues. Figure 10a shows a 14 slots ATCA shelf with a height of 13U, and figure 10b shows a processor ATCA module.

a

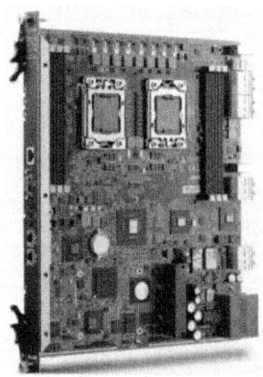

b

Figure 10: a) ATCA shelf of 14 slot 13U, b) Processor ATCA module.

ATCA Modules

Regarding the ATCA modules we can highlight three main types of ATCA blades for the data transport purposes: Front boards, Rear Transition Modules (RTM) and Advance Mezzanine Cards (AMC). All of these modules are hot swappable and have different form factors.

- Front boards are connected to the shelf backplane through the Zone 1 and Zone 2 connectors. Zone 1 connector is use to feed the module and Zone 2 connector is use for data transport signals. Moreover Front boards have a third connector, Zone 3, which provides a direct connection with the RTM.

- Rear Transition Modules are placed in the rear side of the shelf and it is used to expand the ATCA system functionalities.
- AMCs are mezzanine modules pluggable onto ATCA carriers enlarging system functionalities. Examples of AMCs include CPUs, DSP systems or storage.

Moreover, ATCA Fabric interface reaches data rates between modules of 40 Gbps using protocols such as Gigabit Ethernet, Infiniband, Serial Rapid IO or PCIe and network topologies such as dual star, dual-dual star or full mesh. These features provide a clear advantage of ATCA systems over other platforms.

MicroTCA

MicroTCA is a complementary specification to the PICMG 3.0 introduced by PICMG in 2006. It was defined to develop systems that require lower performance, availability and reliability than ATCA systems, and also lower space and cost, but maintaining many features from PICMG 3.0 such as shelf management or fabric interconnects [29].

MicroTCA Shelf and Modules

The shelf can allocate and manage up to 12 single or double size AMCs. AMCs are directly plugged into the backplane in a similar way than ATCA carriers. The function of the backplane is to provide power to the AMC boards and also connection with the data, control, system management, synchronization clock and JTAG test lines. The MicroTCA backplane can implement network topologies like star, dual star, mesh or point-to-point between AMCs. The protocols used for data communications in the MicroTCA backplane are: Ethernet 1000BASE-BX, SATA/SAS, PCIe, Serial Rapid IO or 10GBase-BX4. Data transfers between AMCs within the MicroTCA backplane can achieve speeds of 40 Gbps.

Transmission Media

In the past, copper wires, as coaxial or twisted cables, were widely used to communicate front-end electronics and back-end electronics, or even modules in back-end systems. For example, many NIM and VME modules use coaxial cables with BNC, LEMO or SMA connectors for control or data communications.

But, nowadays, data transmission media for DAQ has moved to fiber optics due to the advantages of fiber optics over copper cables that make them the best option to transmit data in present particle physics experiments. Some of these features are: EMI immunity, lower attenuation, no electrical discharges, short circuits, ground loops or crosstalk, resistance to nuclear radiation and high temperatures, lower weight and higher bandwidth [30].

Due to the widespread use of fiber optics, optic modules play an essential role in present particle physics experiments. Optic modules are needed to convert electrical signals into optical ones for transmission via optic fibers. Some examples of optic modules that are used for data transmission in particle physics experiments and their data bandwidths are shown in the table 3 [31].

Table 3: Examples of parallel optic modules used in particle physics experiments

Type	Channels count/fiber ribbon	Bandwidth (max)
X2	1	10Gbps
XENPAK	1	10Gbps
SFP+	1	10Gbps
XFP	1	10Gbps
QSFP+	4	40Gbps
CFP	10	100Gbps
CXP	12	120Gbps
SNAP12	12	120Gbps
POD	12	120Gbps
MiniPOD	12	120Gbps

BACK-END DATA PROCESSING

In the last decades, the improvements in the analog-to-digital converters, in terms of sampling rate and resolution, opened a wide range of possibilities for the digital data processing. The migration from the analog to the digital processing has proven a number of scenarios where the digital approach has potential advantages, such as system complexity, parameter setup changes or scalability. On the other hand, system designers have to deal with bigger amounts of data processed at higher sampling rates, which affects the complexity of the processing algorithms working in real time and the transport of those data at high rates.

For instance, digital processing has demonstrated significant advantages processing pulses from large-volume germanium detectors, where a good choice in the pulse shaping parameters is crucial for achieving good energy resolution and minimum pulse pile-up for high counting rates.

Common Used Algorithms

Following the inheritance of the analog data processing, some of the digital data processing algorithms perform similar tasks to the analog blocks; taking advantage of the digital information compiled in the ADCs. These algorithms can be divided in five groups:

- Shaping or filtering: When only part of the information from the detector pulses is relevant, such as, for instance, the height of the pulse, shaping techniques can be applied. They filter the digital data according to certain shaping parameters, which could be changed easier in the digital setup. Thus, the only difference in order to apply the same filters in the analog and digital approaches would remain in the continuous or discrete characteristics that differentiate them. In addition, apart from the time-invariant filters similar to the analog ones, in the digital world also adaptive filtering could be applied, changing the filter characteristics for a certain period of time.

- Pulse shape analysis: Exploiting the amount of digital information available in the fast digitalization process, different techniques for the analysis of the shape of the pulses can be applied. According to the detector response, these algorithms could be used for obtaining better detector performance or distinguishing between different input particles, as show in Figure 11.

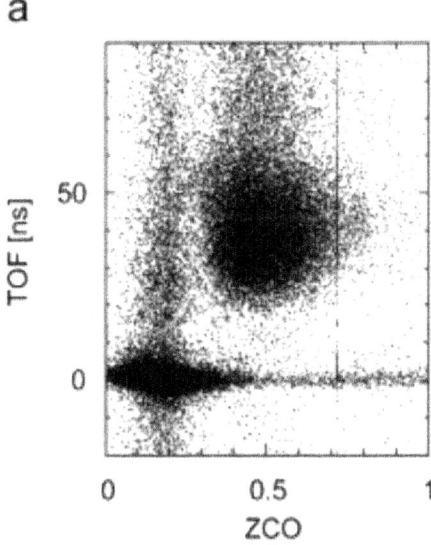

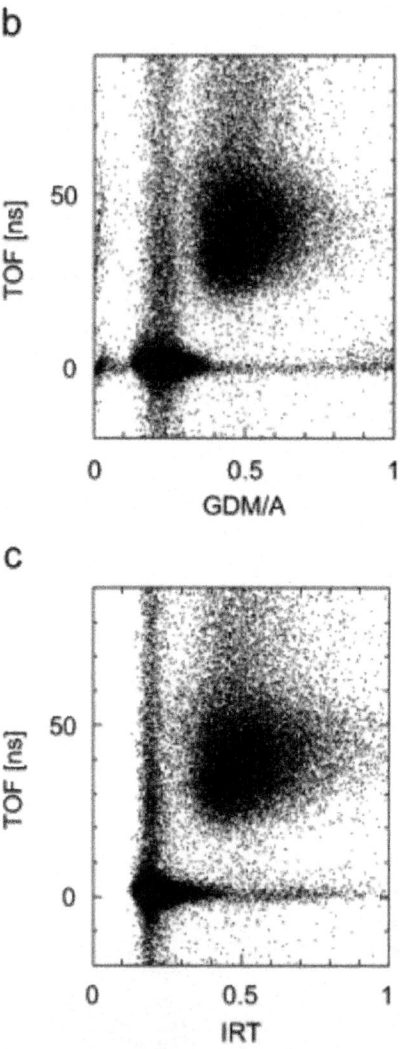

Figure 11: Comparison between an analog (left) and two digital algorithms in a neutron-gamma separation using Pulse shape analysis algorithms [32].

- Baseline restoration: During the time gap between two consecutive pulses, the baseline value can also be digitized and easily subtracted from the digital values of the waveform. Also sometimes more elaborated algorithms can be applied in order to calculate the baseline of the pulses. Thus, better system performance is achieved, avoiding changes due to temperature drifts or other external agents.

- Pile-up deconvolution: The pile-up effect consists on the accumulation of pulses from different events in a short time, which, in principle, avoids the study of those events. In the analog electronics, this effect usually causes an increment in the dead time of the system, as those events should be rejected. However, taking advantage of the digital characteristics, a further analysis of these pulses can be performed and, consequently, in some cases the information of the compound pulses can be disentangled.

- Timing measurements: Timing information is mainly managed in two ways:

- Trigger generation: In a fully digital system, pulse information can be used for generating logical signals for validating certain events of interest. Furthermore, in complex systems with different trigger levels the generation of logic signals for the validation or rejection of events becomes very important. For this purpose, two methods are commonly used: leading edge triggering and constant fraction timing. The first is the simplest, and generates the logic pulse performing a comparison between the input pulse and a constant trigger level. In the second, a small algorithm generates the logic pulse from a constant percentage of the pulse height. There are other algorithms like the crossover timing, ARC timing, ELET, etc. but its usage is lower.

- Measurement of timing properties: With the trigger information generated either analog or digitally, several logic setups can be performed. Thus, depending on the complexity of the experiment and its own characteristics, trigger pulses can be used for measuring absolute timing between detectors, perform new trigger levels according to certain conditions or filter events according to a specific coincidence trigger.

Although the needs of the experiments and the complexity of the setups changes enormously, the processing algorithms usually can fit into one of the categories described previously. However, it is also common to combine several algorithms in the process, so the system architecture can be quite elaborated. The split of the system into several firmware and software blocks allows designers and programmers to manage the difficulty of the experiment.

Hardware Choices

The complexity and performance of the algorithms previously presented varies depending on the input data, the sampling rate or the implementation architecture. For the last one, several options are presented according to the experiment needs:

- Digital Signal Processing (DSP): The DSPs are Integrated Circuits (ICs) that perform programmable filtering algorithms by applying the Multiply-Accumulate (MAC) operation. These devices have been used since more than 30 years, especially in other fields, such as audio, image or biomedical signal processing. The wide knowledge acquired with these devices has contributed to its use within the particle experiments network, taking advantage of its compactness or stability with temperature changes.

- Field-Programmable Gate Array (FPGA): These programmable ICs are composed by a large number of gates that can be individually programmed and linked, as it is depicted in figure 12. They are usually programmed using Hardware Description Languages (HDLs), like the Application-Specific Integrated Circuits (ASICs), so, in theory, they could implement any of the algorithms presented previously. Furthermore, these devices use to embed other DSPs or microprocessors, which allow performing different processing tasks, even concurrently.

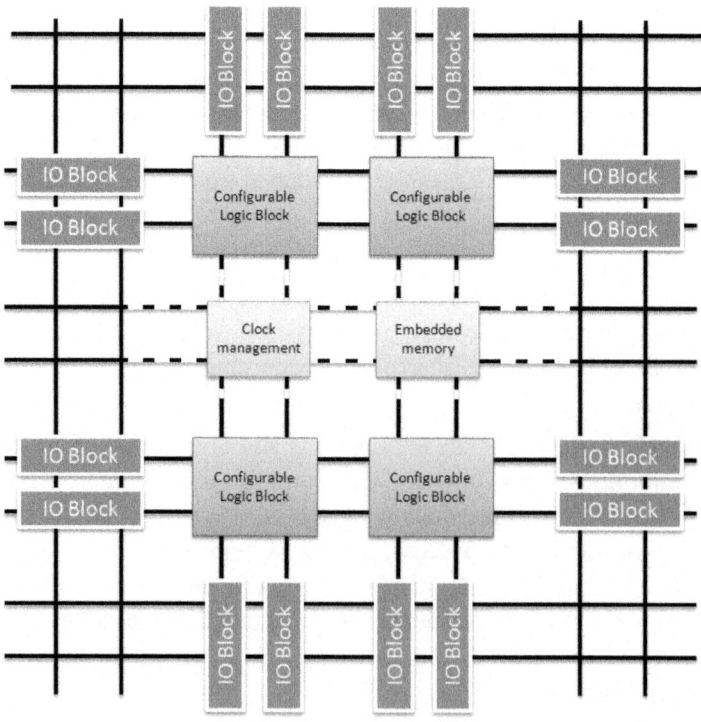

Figure 12: Block structure of a FPGA. It contains the main Input Output Blocks, the Configurable Logic Blocks and other embedded hardware blocks such as Clock management blocks or Embedded memory (SRAM or BRAM based) blocks.

- Central Processing Unit (CPU): In some cases, a Personal Computer (PC) with specific memory or CPU characteristics is used. In this case, the data are treated with software programs together with the operating system. They are often included as an interface to long-term memories, i.e. to handle the storing process. However, sometimes additional data processing is required, and they are particularly efficient when there is a high amount of data, which are not needed to be processed at a high frequency. Moreover, data processing at this point sometimes requires a lot of computing resources, so these processes use to run in computer farms, which are handled by distributed applications or operating systems.

- Graphics Processing Unit (GPU): When parallel processing and the amount of data to manage increase, CPUs may not be the best hardware architecture to support it. Thus, the underlying idea of the GPUs is to use a Graphical card as a processing unit. They have been shown as very efficient hardware setups that can be more efficient than CPUs in some configurations.

- Grid computing: This technique, which is used in large experiments where the amount of data is not even manageable by computing processing farms, consists of taking advantage of the internet network to communicate computing farms or PCs in order to perform a large data processing with an enormous number of heterogeneous processing units. Obviously this technique is implemented with data without timing constraints, as the processing time for each unit may be different.

Among this short description of data processing hardware blocks, an overview of all possible units has been presented. However, it is important to remark that the final system requirements may advise to select a certain setup. For this purpose, the reader is encouraged to review the bibliography for further details.

APPLICATION EXAMPLES

This section reviews at a glance some implementations of DAQ in particle physics experiments and medical applications in order to clarify the concepts on detectors, algorithms and hardware units previously described.

The first example regards to the data processing in the Advanced GAmma Tracking Array (AGATA) [33]. As it is a "triggerless" system, the detector signals (HPGe crystals) are continuously digitized and sent to the pre-processing electronics, which implement shaping, baseline restoration, pile-up deconvolution and trigger generation algorithms in FPGAs in a fully digital way. After that, crossing different bus domains, the data arrive to a PC that performs Pulse Shape Analysis algorithms to calculate the position of the interaction and add it to the energy information calculated in the preprocessing. Then, data coming from all detectors arrive to a PC (Event Builder), using a distributed Digital Acquisition program. When the event is generated, its information is added to the data from other detectors (ancillaries) in a PC called "Merger", that sends them to the tracking processor. This is another PC that performs tracking algorithms for reconstructing the path of the gamma-rays in the detectors. Finally, the data is stored in external servers.

In this example, most of the presented algorithms and hardware configurations are used. In addition, GPUs have been tested for the Pulse Shape Analysis and they have shown excellent performance characteristics. Also Grid computing techniques are used for data analysis and storage, so the hardware components previously presented are almost covered in this example.

The second example is the data processing in the ATLAS detector, at CERN [34]. In this case, the system is composed of several different types of sensors and the DAQ system is controlled by a three level trigger system. In the first level of trigger, the events of interest recorded in the detectors, mostly selected by comparators, are directly sent to the second level of trigger. In this level, the information from several sub-detectors is correlated and merged according to the experimental conditions. Finally, a third event filtering is carried out with the data from the whole system. Along these trigger levels, different processing algorithms are used, combined in different hardware setups. However, all of them can be included in one of the categories previously detailed. An example of the hardware systems developed for this particle physics experiment may be found in [35, 36].

The last example is from radiation therapy where ionizing radiation is used with medical purposes. Nowadays, radiologists make use of radioactive beams i.e. gamma particles, neutrons, carbon ions, electrons, etc. to treat cancer, but they also take advantage of the properties of the ionizing radiation and its application in the diagnosis of internal diseases through medical imaging. This field has involved the development of devices capable of, on one hand, producing the radioactive particle needed for a specific treatment and, on the other hand, to detect the radioactive beam (in some cases, the part of the radiation that has not been absorbed by the patient) to reconstruct the internal image. This is the case of the Computed Tomography (CT) scan, a medical imaging technique consistent on an X-Ray source (X-Ray tube) that rotates 360° around the patient providing at each rotation a 2D cross-sectional image or even a 3D image by putting all the scans together through computing techniques. The detection of the X-Rays is carried out whether in a direct or in an indirect way depending on the device; actually, the detection area consists from one up to 2600 detectors of two categories, scintillators (coupled to PMTs) or gas detectors. Another well-known imaging technique is the Positron Emission Tomography (PET). It provides a picture of the metabolic activity of the body thanks to the detection of the two gamma rays that are emitted after the positron annihilation produced by a radionuclide previously inserted into the patient. Gamma detection is achieved by placing scintillators generally coupled to PMTs but also to Si APDs.

CONCLUSION

In this chapter we have presented a review of the technologies currently used in particle physics experiments following the natural path of the signals from the detector to the data processing. Even these kinds of applications are well established there is not a comprehensive review, as this chapter tries in a very light version, of the overall technologies commonly in use. Being a wide field, we have tried to be concise and provide the interested reader with a list of references to consult.

ACKNOWLEDGEMENT

This work is funded by Spanish Commission of Science and Technology under project reference FPA2009-13234-C04-02

REFERENCES

1.   S. S. Iyengar, et al.1995Advances in distributed sensor integration. Application and theory. Prentice Hall, 1995.

2.   R. Wesson, et al.1981Network structures for distributed situation assessment. IEEE Trans. on System, Man, Cybernetics. 11523

3.   V. González, E. Sanchis, G. Torralba, J. Martos, 2002Comparison of parallel versus hierarchical systems for data processing in distributed sensor networks. IEEE Transactions on Nuclear Science, 49394400

4.   V. González, E. Sanchis, 2004Comparison of data processing techniques in sensor networks, In: Ilyas Mahgoub, editor. Handbook of Sensor Networks: compact wireless and wired sensing systems. CRC Press. 21

5.   N. Tsoulfanidis, 1995Measurement and detection of radiation (2nd Edition). Taylor & Francis.

6.   W. R. Leo, 1994Techniques for Nuclear and particle Physics Experiments (2nd Edition). Springer-Verlag.

7. G. F. Knoll, 2000Radiation Detection and Measurement (4th Edition). John Wiley & Sons.

8. Kaufmann K.J.2000Light levels and noise. Guide detectors choice. Photonics Spectra, 7.

9. Kaufmann K.J.2005Choosing your detector. OE Magazine.

10. D. Renker, 2006Geiger Mode avalanche photodiodes, history, properties and problems. Nuclear Instruments and Methods in Physics Research A, 5674856

11. P. Hamamatsu, 2008Characteristics and use of Si APD (Avalanche Photodiode) Technical information, SD-28.

12. P. Hamamatsu, 2009Photodiode Technical Information. Application note, SD-03.

13. T. J. Sobering, 1999Bandwidth and rise time. Technote 2, SDE Consulting. May 1999.

14. P. Hamamatsu, 2001Characteristics and use of the Charge Preamplifier, Application note SD-37, 2001.

15. A. Rivetti, et al.2011Integrated circuits for time and amplitude measurement of nuclear radiation pulses. Short course on IEEE 2011 Nuclear Science Symposium and Medical Imaging Conference, Valencia, October 2011.

16. Analog Devices.2006High Speed System Applications.

17. W. Kester, 2004The data conversion handbook, Analog Devices.

18. Hofner T.C.2002Measuring and evaluating dynamic ADC Parameters. Microwaves & RF, 7894

19. M. Joos, 2012Standards for Modular Electronics- the past, the present and the future. International School of Trigger and Data Acquisition 2012, Cracow, Poland.

20. S. N. Ahmed, 2007Physics and Engineering of Radiation Detection. Academic Press, Elsevier.

21. VITA Technologies2011VME Technology FAQ. Available: http://www.vita.com/home/Learn/vmefaq/vmefaq.html.Accessed 2012 May 10.

22. VITA Technologies.Frequently asked questions on VXS. Available: http://www.vita.com/home/MarketingAlliances/vxs/vxs_faq.pdf.Accessed 2012May 10.

23. PCI-SIG.2002PCI Local Bus Specification Revision 2.3, PCI-SIG.

24. Digi International2003PCI Technology Overview. Available: http://www.digi.com/pdf/prd_msc_pcitech.pdf.Accessed 2012 May 10.

25. M. Matveev, 2008Databus Systems in High Energy Physics (HEP): Past, Present and Future, Rice University, Houston, Texas, USA.

26. PCI-SIG.2003PCI Express Base Specification Revision 1.0a.

27. A. H. Wilen, J. P. Schade, R. Thornburg, 2003Introduction to PCI Express. A Hardware and Software Developer's Guide, Intel Press.

28. PICMG.2003PICMG 3.0Advanced TCA Short Form Specification, Available: http://www.picmg.org/pdf/picmg_3_0_shortform.pdf. Accessed 2012 May 10.

29. S. Jamieson, 2006PICMG MTCA.0 R1.0Micro Telecommunications Computing Architecture Short Form Specification. Available:http://www. picmg.org/pdf/microtca_short_form_sept_2006.pdf.Accessed 2012 May 10.

30. A. A. Gokhale, 2004Introduction to Telecommunications, Cengage Learning.

31. J. Chramowicz, S. Kwan, A. Prosser, M. Winchell, 2011Application of emerging parallel optical link technology to high energy physics experiments, Proceeding of 2nd International Conference on Technology and Instrumentation in Particle Physics 2011, Chicago.

32. P. A. Söderström, J. Nyberg, R. Wolters, 2008Digital pulse-shape discrimination of fast neutrons and rays. Nuclear Instruments and Methods in Physics Research Section A: Accelerators, Spectrometers, Detector and Associated Equipment, 5947989

33. S. Akkoyun, et al.2012AGATA- Advanced Gamma Tracking Array. Nuclear Instruments and Methods in Physics Research Section A: Accelerators, Spectrometers, Detector and Associated Equipment, 6682658

34. ATLAS Collaboration.1994ATLAS Technical Proposal for a General-Purpose pp Experiment at the Large Hadron Collider at CERN, CERN, Switzerland.

35. V. González, E. Sanchis, et al.2006Development of the Optical Multiplexer Board Prototype for Data Acquisition in the TileCal System. IEEE Transactions on Nuclear Science, 5321312138

36. F. Carrió, V. Castillo, V. Ferrer, V. González, E. Higón, C. Marín, P. Moreno, E. Sanchis, C. Solans, A. Valero, J. Valls, 2011Optical link card design for the upgrade phase II of TileCal experiment. IEEE Transactions on Nuclear Science, 5816571663

Chapter 4

ON THE NEED FOR FRACTAL LOGIC IN HIGH ENERGY QUANTUM PHYSICS

M. S. El Naschie[1], S. Olsen[2], J. H. He[3], S. Nada[4], L. Marek-Crnjac[5], A. Helal[6]

[1]Department of Physics, Faculty of Science, University of Alexandria, Alexandria, Egypt

[2]College of Central Florida, Ocala, USA

[3]Nantong Textile Institute, National Engineering Laboratory for Modern Silk, College of Textile and Clothing, Soochow University, Suzhou, China

[4]Mathematics Department, Qatar University, Doha, Qatar

[5]Technical School Center of Maribor, Maribor, Slovenia

[6]Department of Mathematics, Faculty of Science, Cairo University, Giza, Egypt

ABSTRACT

Modern advances in pure mathematics and particularly in transfinite set theory have introduced into the fundamentals of theoretical physics many novel concepts and devices such as fractal quasi manifolds with non-integer (Hausdorff) dimension for its geometry as well as infinite dimensional wild topology and non classical fuzzy logic. In the present work transfinite fractal sets and fuzzy logic are combined to enable the introduction of a new theory termed fractal logic to the foundation of high energy particle physics. This leads naturally to a new look at quantum gravity. In particular we will show that to understand and develop quantum gravity we have to bring various fields together, particularly fractals and nonlinear dynamics as well as sphere packing, fuzzy set theory, number theory and quantum entanglement and irrationally q-deformed algebra.

INTRODUCTION AND BACKGROUND INFORMATION

There is a large body of literature on the application of fractals and deterministic chaos in quantum physics [1-69]. In particular, the present authors integrated Cantor sets and fractals as well as number theory into the foundation of quantum mechanics [5-22]. Modern transfinite set theory [6] has unraveled many undreamed of logical possibilities such as hierarchies of things and "numbers" larger than infinity [25] and smaller than zero in addition to non integer and even negative dimensions [1,2,4,5,7,8,10]. Some of these novel and particularly revolutionary mathematical notions have sometimes ago found their way into theoretical physics [1-16]. This is definitely the case with fractal-Hausdorff dimensions [11, 14,15] which led to the theory of fractal space time pioneered by G. Ord [7]. Other even more demanding concepts such as the empty set and negative topological dimensions form the basis of El Naschie's Cantorian E-Infinity theory [5,10,31,32]. The present note continues this line of thinking by pointing out the urgent need for a new theory of general fractal logic in high energy quantum physics when counting and discovering new particles which is an essential part of the trade. The idea is related but not identical to Zadeh's theory of fuzzy logic [33-38]. Our present main thrust is to reason that although counting which is at the basis of statistical mechanics may seem like the simplest thing in the universe, high energy physics calls for another form of fractal transfinite counting which is anything but straightforward in the traditional sense [5-10]. As a clear cut example of profound consequence we show that while the number of gauge bosons of the standard model is indeed 12 in the ordinary sense of counting using integer numbers [55-57], these 12 corresponding to only 11.7082394 bosons which have in turn the fractal weight of 14 rather than 12 gauge bosons [38-54]. In this sense we are asserting what on its face value is an absurdity namely the equality of 12 to 14 via 11.708294. This indicates that 2 more elementary particles are hidden allowing 14 particles to appear as if they were 12. We conjecture that these two extra particles may be the Higgs and the graviton. It is further concluded that this fractal nature which is inert even to the mere number of elementary quantum particles is in one way or another behind the enormous difficulties of discovering the Higgs and the graviton experimentally [55-57]. This is more than sufficient reason to call for the introduction of a new theory for fractal logic.

RADON WORK AND TRANSFINITE HARDY QUANTUM ENTANGLEMENT

In order to be able to analytically observe the transfinite irrational fingerprint of the golden mean entanglement [65-67], we must fulfill the following conditions:

- First we must have an accurate mathematical model which can capture a substantial and relevant part of Reality.
- Second we must solve the model exactly.

If we do not observe both of the above conditions, the golden mean fine structure, i.e. the fractal Cantorian structure disappears theoretically and would unlikely be observed experimentally [52].

Adding or subtracting small quantities with the objecttive of simplifying the corresponding field equation is a method which goes back to the work of F. John on the derivation of the equations of thin elastic shells using Taylor expansion from the 3D field equations of the classical theory of Elasticity [63]. Interestingly John based his method on the work of Radon and his theory of operators. In turn Radon's work was extended by Von Neumann and later on by A. Connes in his theory of non-commutative geometry [10-63]. Finally all of this is strongly related to El Naschie's E-Infinity theory [10,63]. It is thus understandable that E-Infinity extended and made extensive use of F. John's method to make it applicable to quantum high energy physics [52,59]. The astonishing point is that the so-called transfinite correction corresponding to F. John's method is almost entirely reducible to Hardy's generic value $f^5 = 0.09016994393$ which is the quantum probability of Hardy's entanglement [30] where
$$\phi = \frac{\sqrt{5}-1}{2}$$

is the inverse of the golden mean. For instance $\bar{\alpha}_0$ is simply the prime number 137 plus a function of f^5 namely $k_0 = f^5(1 - f^5)$. We may thus write that [10,31,32]

$$\bar{\alpha}_0 = 137 + k_0$$
$$= 137 + \phi^5 \left(1 - \phi^5\right)$$
$$= 137.082039325 \qquad (1)$$

where $\bar{\alpha}_0$ is the inverse electromagnetic fine structure constant [56,58].

Now Measure Theory [29] gives us the tools to estimate the quantities needed to make things and equations "fit" harmonically together. In a manner of speaking transfinite corrections a la F. John and El Naschie is developed to make the rough or sharp edges of the equation and their geometrical shapes fit together like in Penrose's golden mean fractal tiling [1] or a bathroom floor where transfinite "Mortel" fill the gaps between the edges the tiling to make for a smooth tiling similar to what we have in "Al Hambra" Islamic wallpaper groups, the 2 and 3 Stein spaces and the compactified Kleinian curve SL(2, 7) used for the Holographic boundary of E-Infinity [1-73]. Meantime it is well understood that this Holographic boundary manifold [2] is related to the bulk of E_8E_8 exceptional symmetry group via the fundamental equation B = H + G + E where B is the Bulk, H is the Holographic boundary, G is Einstein gravity and E is Electromagnetism. Since H = 339, G = 20 and E = $\bar{\alpha}_0$ =137 one finds the integer value of B to be $B = 496 = E_8E_8$ exactly as should be [20,63]. However the exact transfinite-corrected value of B is found directly from the corresponding transfinite value of $\bar{\alpha}_0 = 137 + k_0$ by the exact golden mean scaling which replaces the classical approximate logarithmic scaling of conventional quantum field theory [62]. The golden mean scaling exponent in this case is $\lambda = 3 + f = 3.618033989$ so that the exact B is found to be [10,63].

$$
\begin{aligned}
B &= |E_8E_8| \\
&= (\bar{\alpha}_0)(\lambda) \\
&= (137 + k_0)(3 + \phi) \\
&= 496 - k^2
\end{aligned}
\tag{2}
$$

where $k = 2f^5 = f^3(1 - f^3)$ and f^5 is the probability of Hardy's quantum entanglement [30].

The preceding result may be interpreted in terms of quasi particles as follows: We have an expectation average KAON [10] with mass equal 496 GeV meeting a virtual particle with negative mass equal $-k^2 = -0.032522475$ GeV and producing $496 - k^2$ which in this unit-less form is the quasi dimension of the bulk. A similar interpretation may be given to $\bar{\alpha}_0 = 137 + k_0$ [10,63].

We may also note a very nice symmetry or rather anti-symmetry between the transfinite correction of E_8E_8 and that of $\bar{\alpha}_0$ and $\sqrt{\bar{\alpha}_0}$.

This is obvious from contrasting k and k_0 and remembering that, we subtract k^2 from the exceptional 496 symmetry group dimension while we add k_0 to the 137 of alfa:

$$
\begin{aligned}
\sqrt{\bar{\alpha}_0} &= 12 - 2\phi^4 \\
&= 12 - \left[\phi^3 \left(1 + \phi^3 \right) \right]
\end{aligned}
\tag{3a}
$$

and

$$
k_0 = \phi^5 \left(1 - \phi^5 \right), k = \phi^3 \left(1 - \phi^3 \right).
\tag{3b}
$$

We recall that f^5 is the contra factual or global probability of quantum entanglement [30-32] which together with the local two-particle quantum entanglement f^2 gives us the Hardy generic value of the probability of quantum entanglement namely $f^5 = (f^3)(f^2)$. For readings on the method of adding and subtracting small quantities to smooth the given equation of classical field theories, we refer to the work of F. John and M. S. El Naschie [10,60-63].

SPHERE PACKING AND GOLDEN MEAN TRANSFINITE-NESS

The connection between "fractal" sphere packing and E-Infinity theory was considered in detail by L. MarekCrnjac and El Naschie [2,10]. Here we recall some remarkable observation which serves to deepen our understanding of transfinite golden mean corrections [30-32].

- First we note that the density of sphere packing in four dimensions is Δ = 0.61685. This is very close to the golden mean f = 0.618033989.

- In the case of ten dimensions i.e. the dimensionality of superstring theory, the lattice sphere packing has a density of Δ = 0.09202 $\approx$ 9%.

  This is rather close to Hardy's quantum entanglement probability.

- Curiously, at this ten (10) dimensionality the central density is 0.03608. This is very close indeed to $k_0(\exp)$ = 0.0359 which appears in the experimental determination of $\bar{\alpha}_0$

$$\begin{aligned}
\bar{\alpha}_0 &= 137 + k_0(\exp) \\
&= 137 + 0.0359 \\
&= \frac{137}{\cos \pi} = 137.0359852 \\
&\quad \bar{\alpha}_0
\end{aligned}$$

(4)

We note that the theoretical value of k_0 is $k_0 = f^5(1 - f^5)$ where f^5 is Hardy's quantum entanglement probability [30].

WHEN THE 12 BOSONS OF THE STANDARD MODEL BECOME EQUIVALENT TO THE 14 BOSONS OF SUPER GRAVITY VIA THE NEW THEORY OF FRACTAL LOGIC

The classical standard model which is at the core of conventional quantum field theory relies and stems from 12 massless gauge bosons described by a 12 dimensional combined Lie symmetry group SU(3)SU(2)U(1). Apart from superstring theory which proceeds in an entirely different way with 496 massless bosons corresponding to the generator of $E_8 E_8$ exceptional Lie group, there is a more conventional approach which is based on an eleven dimensional space-time similar to the acclaimed eleven dimensional M-theory of E. Witten [49,50]. This theory is known as super gravity and relies neither on $E_8 E_8$ nor on SU(3)SU(2)SU(1) but rather on a less known group with 14 dimensions [55]. This is the ortho-symplectic group OSP(m/n) where [55,56] exactly as anticipated [55].

$$DimOSP(m/n) = \frac{1}{2}\left[(m+n)^2 + (m-n)\right]$$

(5a)

where

$$DimOSP(1/4) = \frac{1}{2}\left[(4+1)^2 + (4-1)\right]$$
$$= \frac{1}{2}[25+3]$$
$$= 14$$

(5b)

In what follows we will attempt to reason that the 12 bosons of the standard model are in reality 14 by virtue of a fractal number

$$\sqrt{\bar{\alpha}_0} = 11.708203932$$

(6)

which give a real fractal weight to 12 bosons and make them equivalent to 14 in this fractal counting of fractal elementary particles using what we may call fractal logic. This is a notion related but not identical to fuzzy logic as mentioned earlier on [33,38].

Our next and most important novel task is to devise a fractal number to massless gauge bosons of the standard model. There is no automatic mechanistic way to do that at present but we are guided by intelligent guess work and the overriding principle of a harmonic number theoretical symphony which fits everything with everything else seamlessly [9,65]. This principle served us well in the past and as we will see shortly will lead us to also here astonishing undreamed of possibilities coupled with the utmost simplicity [36-53].

The first and may be the most important messenger particle of them all, the illusive photon (γ) will be given the fractal number weight not 1 but f. This f we recall is at the same time the Hausdorff dimension of a quantum particle whose cobordism [20,30] namely the quantum wave was given the Hausdorff dimension f^2 [20]. We also recall that f corresponds to the zero set while f^2 corresponds to the classical empty set with the MengerUhryson topological dimensions zero and minus 1 respectively [10,61].

Second the three bosons of the electro weak are given a priori an easily justifiable weighted number namely $16k = 32f^5$. Third the 8 gluons are given the fractal weight number $8 + (^{k/2}) = 8 + f^5$. The total is now 11.5936411 weighed fractal particles corresponding to 12 particles. To this we add two new particles which we conjecture that they could be one Higgs and a graviton. The fractal number of the Higgs is $k_0 = f^5(1 - f^5)$ while that of the graviton is $k^2 =$

$[f^5(1 - f^5)]^2$. The total is now 11.70820393 which happen to be $\sqrt{\bar{\alpha}_0}$ where $\bar{\alpha}_0 = 137 + k_0$ is the E-Infinity value of the inverse electromagnetic fine structure constant [10,63].

The $\sqrt{\bar{\alpha}_0}$ massless bosons correspond therefore to 14 particles when counting in the ordinary way using integers. Consequently they could be represented by the OSP(1/4) symmetry group of super gravity [55] and it is granted that this astonishing unexpected fact gives a feel similar to that of Alice in the Fractal Land of high-energy quantum physics [10,30-32]. The preceding result may be summarized in the following table:

Standard Model		Fractal fuzzy logic
1 Photon γ. $\|U(1)\| = 1$	$\longrightarrow$	$\phi = \dfrac{\sqrt{5} - 1}{2}$
3 $W^+.W^-.Z^0$ $\|SU\| = 3$	$\longrightarrow$	$16k = 32\phi^5$
8 Gluons $\|SU(3)\| = 8$	$\longrightarrow$	$8 + (k/2) = 8 + \phi^5$
1 Higgs = 1	$\longrightarrow$	$k^2 = \left[\phi^3\left(1 - \phi^3\right)\right]^2$
1 Graviton = 1	$\longrightarrow$	$k_0 = \phi^5\left(1 - \phi^5\right)$
Sum 12 + 2 = 14	$\longrightarrow$	11.708203932

THE LOGARITHMIC SCALING OF QUANTUM FIELD THEORY AS AN APPROXIMATION OF THE GOLDEN MEAN SCALING OF E-INFINITY

The discovery of the logarithmic scaling of high energy physics was a major step forward for establishing quantum field theory [56,57]. Nonetheless logarithmic scaling as will be shown here is merely an excellent approximation of the exact golden mean scaling of E-Infinity golden field theory. As excellent as it is, this approximation masks Hardy's golden mean quantum entanglement and makes algebraic manipulation of the governing equations unnecessarily heavy if not clumsy and sometime interactable and messy [55-57]. As a simple demonstration testifying for the above, let us consider the logarithmic scaling of the inverse fine structure constant of a Cooper Pair. This is clearly

$$\ln\left(\frac{\bar{\alpha}_0}{2}\right) = 4.2274323923$$

[58-64]. By comparison the golden mean scaling for $\dfrac{\bar{\alpha}_0}{2}$ namely $\lambda = \dfrac{\phi}{10}$ leads to

$$\lambda\left(\frac{\bar{\alpha}_0}{2}\right) = \left(\frac{\phi}{10}\right)\left(\frac{\bar{\alpha}_0}{2}\right) = 4.236067977 = 4 + \phi^3$$

(7)

which is equal to the exact Hausdorff dimension of the core of E-Infinity Cantorian space-time.

The above result must make us wonder why not reconsider the famous renormalization equation of quantum gravity unification using golden mean scaling. Let us write the concerned renormalization equation in the logarithmic scaling form first. This is [61,62]

$$\bar{\alpha}_u = \bar{\alpha}_3 + \bar{\alpha}_4 + \frac{1}{\delta}\ln\left(\frac{M_u}{M_x}\right)$$

(8)

Let us replace in this equation the logarithmic scaling term by the golden mean scaling as follows [60-62]

$$\ln\left(\frac{M_u}{M_x}\right) \Rightarrow (\bar{\alpha}_0)(\phi^3)$$

(9)

Then insert all the exact E-Infinity value of $\bar{\alpha}_3 = 9$ and $\bar{\alpha}_4 = 1$, $\delta = \dfrac{1}{2}$ and $\bar{\alpha}_0 = 137.082039$ in the renormalization equation of ordinary quantum field theory. That way we find [60-64]

$$\bar{\alpha}_u = \bar{\alpha}_3 + \bar{\alpha}_4 + \frac{1}{\delta}(\bar{\alpha}_0)(\phi^3)$$

$$= 9 + 1 + \frac{1}{2}(32 + 2k)$$

$$= 10 + (16 + k)$$

$$= 26 + k = 26.18033989$$

(10)

where k = f³(1 − f³).

This is the well known exact E-Infinity result [65,67].

To ascertain the high accuracy of the classical theory based on logarithmic scaling compared to our exact solution, we repeat the analysis and replace $(\bar{\alpha}_0)(\phi^3)$ by what we have derived at the very beginning namely

$$10\left(\ln\frac{\bar{\alpha}_0}{2}\right)\phi^2$$

and find that [61-63]

$$\overline{\alpha}_u = (\overline{\alpha}_3 + \overline{\alpha}_4) + (10)(4.2274)(\phi^2) = 26.12944393 \tag{11}$$

This is pretty close to the exact result of E-Infinity theory namely 26.18033989 [10,43,63]. Of course the above equation could have been performed without using $\overline{\alpha}_0$ nor f^2. We could have used the conventional purely logarithmic running of coupling by taking for instance M_u to be that of grand unification i.e. $M_u (10)^{16}$ GeV while taking $M_x = M_z = 91$ GeV. This leads to [61-63]

$$\ln \frac{(10)^{16} \, \text{GeV}}{91 \, \text{GeV}} = 42.33050198 \tag{12}$$

For non-super symmetric and super symmetric unification respectively, we found the values $\overline{\alpha}_g = 42.33050198$ and $\overline{\alpha}_u = 26.16525099$. Both values are very close to the exact results [48].

TOWARD A THEORY OF FRACTAL LOGIC

Even today some would dispute that Fuzzy logic is a new theory [35]. We beg to differ and insist that Fuzzy logic [33-38] was and is still a radically novel theory with far reaching consequences which has played a significant role not only in running production lines of modern care industries but also in furthering and deepening understanding of fundamental theories such as quantum physics and deterministic chaos [61,62]. In what follows we list a large number of points and observations which when taken together indicate in our opinion that we are on the verge of discovering even more deep logical theories than Fuzzy Logic [33-38] which we have dubbed for the time being Fractal logic for want of a better name. Without much ado here are the points which we gathered for the purpose of the present article.

It is Currently a Widespread Belief that We Have 3 Different High Energy Theories Which Go beyond the Standard Model Namely

- Super gravity [55];
- Technicolor;
- Extra dimensions [55].

However the point is that all the three theories are basically equivalent when pondered via the fractal fuzzy logic of E-Infinity theory [10,37,51]. The reason is simply the fractal fuzzy equivalence between particles, extra dimensions and color [41]. Within E-Infinity these concepts are fractally more or less logically the same [41,42,68]. In fact probability and dimension are in some E-Infinity sense the same [63].

Putting a Photon in a Fractal M-Theory Space-Time Leads to an Extended Standard Model and to $\bar{\alpha}_0 \equiv 137$

In a nutshell this could be explained as follows: A fractal 11D theory has a Hausdorff dimension [49,50,55]

$$D_F^{(11)} = 11 + \cfrac{1}{11 + \cfrac{1}{11 + \cdots}}$$

(13)

where $\phi^5 = \dfrac{1}{D_F^{(11)}}$.

At the same time the fractal weighted number of the $14 = 12 + 2$ particles of OSP(1/4) super gravity is

$$N = \sqrt{\bar{\alpha}_0} = \left(11 + \phi^5\right) + \phi = D_F^{(11)} + \gamma$$

(14)

where the photon γ is given the fractal number

$$\phi = \frac{\sqrt{5} - 1}{2}$$

Consequently $\bar{\alpha}_0$ is given by [49,50,55]

$$\bar{\alpha}_0 = \left(D_F^{(11)} + \gamma\right)\left(D_F^{(11)} + \gamma\right) = \left(D_F^{(11)} + \gamma\right)^2$$

(15)

Noting that $|E_8 E_8|$ is given by $|E_8 E_8| = (3 + f)(\bar{\alpha}_0) = 496 - k^2$ and that

$$\frac{\left(496 - k^2\right)}{\sqrt{\bar{\alpha}_0}} = \bar{\alpha}_g = 42 + 2k$$

(16)

Then by analogy we have $\dfrac{(10)^{19}}{\sqrt{\bar{\alpha}_{Newton}}} = \bar{\alpha}_{QG}$

Here $\bar{\alpha}_{QG}$ is the inverse of quantum gravity coupling which must evidently be unity because it is the coupling of Planck masses to the Planck Aether i.e. to itself and could therefore be maximally equal to unity $\bar{\alpha}_{QG} = 1$. On the other hand $(10)^{19}$ corresponds to the generator of $E_8 E_8$. This is effectively a manifold with $(10)^{19}$ degrees of freedom which may be interpreted as the number of nucleons in a plank mass [67,68]. Since a Plank mass is about $(10)^{19}$ GeV and a nucleon is about 1000 GeV then we have nucleons corresponding to the degrees of freedom mentioned earlier on. The only unknown in the equation is thus $\sqrt{\bar{\alpha}_{Newton}}$ which we inserted by analogy to $\sqrt{\bar{\alpha}_0}$ and which is the number of massless bosons in the standard model. Consequently by solving for $\sqrt{\bar{\alpha}_{\text{Newton}}}$ we find [58-63]

$$\sqrt{\bar{\alpha}_{\text{Newton}}} = \frac{(10)^{19}}{\bar{\alpha}_{QG}} = \frac{(10)^{19}}{1}$$

(17a)

and therefore

$$\bar{\alpha}_{\text{Newton}} = \left[(10)^{19}\right]^2 = (10)^{38}$$

(17b)

This is exactly the right order of magnitude expected for the unitless Newton gravity constant [67,68].

Why Ten Dimensional Superstrings and Why Fractal Eleven Dimensional M-Theory?

There is an excessively simple and elementary "fractal" way to persuade one at least intuitively that three should be an eleven dimensional M-like fractal space-time [49,50].

We know that bosons live in $3 + 1 = 4$ dimensions, while fermions need a spin $1/2$ degree of freedom and need therefore $4 + 1 = 5$ dimensions. Now what is the dimensionality of the space which combines both? A rather naive answer which turns to be the dimensionality of an 8D super space is to take $(4 - 1)$D and add $(5 - 1)$D to it and then add the time dimension and find $(4 - 1) + (5 - 1) + 1 = 7 + 1 = 8$D [55,63].

On the other hand arithmetic mean of the intersection of 4D and 5D gives us the dimensionality of superstrings because

$$D(\text{Superstring}) = \frac{(4)(5)D}{2} = \frac{20D}{2} = 10D$$

(18)

Now if we recall that the Hausdorff dimension of bosonic space is not 4 but $4 + f^3$ and that the Hausdorff dimension of the fermionic space is not 5 but $5 + f^3$ then we will realize that in this case the dimension corresponding to $D(\text{Superstring}) = 10$ is

$$D = \frac{(4+\phi^3)(5+\phi^3)}{2} = \frac{(22.18033988)}{2}$$

$$= 11 + \phi^5 = 11 + \cfrac{1}{11 + \cfrac{1}{11 + \cdots}}$$

(19)

This is the dimensionality of the fractal M-theory as we may have surmised from the outset [49-51].

Quantum Dimension, Irrationally Deformed Algebra and the Golden Oscillator

As well known from the theory of quantum groups, the quantum dimension is given by [46]

$$D = \frac{\left(\frac{1}{q}\right)^{n+1} - q^{n+1}}{\frac{1}{q} - q} \tag{20}$$

On the other hand we know that maximally irrationally deformed algebra for which n = 2 corresponds to two degrees of freedom oscillator which acquire for certain unitary parameters a frequency equal to

$$q = \phi = \frac{\sqrt{5} - 1}{2} .$$

Inserting in D one finds [46]

$$D = \frac{\left(\frac{1}{q}\right)^3 - \phi^3}{\frac{1}{q} - \phi} = \frac{\left(4 + \phi^3\right) - \phi^3}{\left(1 + \phi\right) - \phi} = 4 \tag{21}$$

This is a remarkable result particularly when we recall the Hausdorff dimension of Cantorian space-time is given by [10]

$$\langle d_c \rangle = \frac{\left(1 + \phi\right)}{\left(1 - \phi\right)} = 4 + \phi^3 \tag{22}$$

so that the two dimensions D and $\langle d_c \rangle = \langle n \rangle$ are in dual relations [46]. The result shows indirectly that the golden mean number system used in our E-Infinity and fractal logic is naturally quantized and may be regarded as a generalization of the machinery of quantum groups [46]. This realization may be used advantageously in quantum mechanics and high energy physics as done in E-Infinity theory [10,63].

An Infinite Hierarchy of Nothing

Research in set theory has revealed since sometime that there is a hierarchy of infinites with some that are larger than others [6,25]. In this section we show that there is an infinite hierarchy of empty sets i.e. Nothingness [62]. For this demonstration we use the formalism developed in non-commutative geometry of A. Connes which is in full agreement with the formalism and results of E-Infinity theory [63].

We start from the non-commutative geometry dimensional fraction of A. Connes namely [20,31,32]

$$D_n = a + b\phi;\ a, b \in Z;\ \phi = \frac{\sqrt{5}-1}{2}.$$

(23a)

Thus we may write:

$$D_0 = 0 + (1)(\phi) = \phi = d_c^{(0)}$$

$$D_1 = 1 + (0)(\phi) = 1 = d_c^{(1)}$$

$$D_2 = 1 + (1)(\phi) = \frac{1}{\phi} = d_c^{(2)}$$

$$D_3 = 2 + (1)(\phi) = \left(\frac{1}{\phi}\right)^2 = d_c^{(3)}$$

$$D_4 = 3 + (2)(\phi) = \left(\frac{1}{\phi}\right)^3 = d_c^{(4)}$$

$$\vdots$$

$$D_n = a + b\phi = \left(\frac{1}{\phi}\right)^{n-1} = d_c^{(n)}$$

(23b)

where a, b represents classical Fibonacci numbers.

Consequently we have

$$d_c^{(-1)} = \phi^2 \rightarrow \text{empty set}$$

$$d_c^{(-2)} = \phi^3 \rightarrow \text{empty set}$$

$$d_c^{(-3)} = \phi^4 \rightarrow \text{even emptier—still set}$$

$$d_c^{(-4)} = \phi^5 \rightarrow \text{the empty set of Hardy's quantum ent.}$$

$$\vdots$$

$$d_c^{(-n)} = \phi^{n+1}$$

$$\vdots$$

$$d_c^{(-\infty)} = 0 \rightarrow \text{the total empty set or initial sin gularity}$$

(24)

We note that while $d_c^{(0)} = \phi$ represents in E-Infinity theory a quantum particle, the empty set $d_c^{(-1)} = \phi^2$ represents the quantum wave or the surface of the particle [20,31,32].

DISCUSSION AND CONCLUSIONS

The key to understanding the superficial inconsistency coupled with undreamed of possibilities which fractal logic and high energy particle physics offers is to understand the dual nature of the geometry and topology of random Cantor sets and its golden mean Hausdorff dimension [63].

Let us start by asking why random Cantor sets? The answer may be found in the following:

- It has a zero measure i.e. a zero length as well as a zero topological dimension and yet it has a non-zero Hausdorff-fractal dimension. In a sense it is both there and not there at the same time [10,63].

- A Cantor set has the cardinality of the continuum yet it is a totally disjointed discretuum [10,63].

- It possesses the golden mean Hausdorff dimension which is the epitome of regularity and harmonic order, yet it is random [10,63].

- Although its Lebesque measure is zero, its Hausdorff measure is 1 so that with certain fine print one could write for it an absurd looking equation namely that zero = one [31,32].

- The geometry of a Cantor set is the reason behind quantum entanglement [30].

The next question to ask when seeking a deeper understanding of fractal logic is why the golden mean? This may be considered as follows:

- It is the real solution for the simplest quadratic equation.

- It is the most irrational number and thus belongs to most stable periodic orbit as shown in KAM theorems of non linear dynamics [45,62].

- It preserves the simplicity of fractal tiling and symmetry groups such Penrose tiling [1].

- It is the Eigen frequency of the simplest two degrees of freedom mechanical oscillators [10,18].

- It has the simplest continued fraction and continued square root expansion representation [10,62,63].

- It is found abundantly in philo texts of plants as well as in art, paintings and music being a bridge between art and science [9,10].

- It is the raison d'etre for Hardy's quantum entanglement.

- Golden mean arithmetic and number system are a naturally quantized calculus with transition to zero at infinity [31,32].

For all of the above reasons and the other fundamental findings presented in the main body of this paper, it seems that fractal logic is unavoidable and in time it will become the standard tool of theoretical scientists who work in high energy particle physics [65]. We have reached a sophisticated level of our mathematical civilization moving from performing computations and counting using numbers to performing computations and counting using fractals and Cantor sets [65-69].

REFERENCES

1. L. Marek-Crnjac, "The Hausdorff Dimension of the Penrose Universe," Physics Research International, Vol. 2011, 2011, Article ID: 874302.

2. L. Marek-Crnjac, "A Short History of Fractal Cantorian Space-Time," Chaos, Solitons & Fractals, Vol. 41, 2009, pp. 2697-2705. doi:10.1016/j. chaos.2008.10.007

3. J. Hocking and G. Young, "Topology," Dover Publishing, New York, 1961.

4. S. L. Lipscomb, "Fractals and Universal Spaces in Dimension Theory," Springer, New York, 2009. doi:10.1007/978-0-387-85494-6

5. M. S. El Naschie, "Complexity Theory Interpretation of High Energy Physics and Elementary Particle Mass Spectrum," In: B. G. Sidharth, Ed., Frontiers of Fundamental Physics, Vol. 3, Universities Press, Hyderbad, 2007, pp. 1-32.

6. M. Heller and W. H. Woodin, "Infinity: New Research Frontiers," Cambridge University Press, Cambridge, 2011.

7. G. N. Ord, "Fractal Space-Time a Geometric Analog of Relativistic Quantum Mechanics," Journal of Physics A, Vol. 16, No. 9, 1983, pp. 1869-1884. doi:10.1088/0305-4470/16/9/012

8. L. Nottale, "Fractal Space-Time and Micro Physics," World Scientific, Singapore City, 1993.

9. A. Stakhov, "The Mathematics of Harmony," World Scientific, Singapore, 2009.doi:10.1142/6635

10. M. S. El Naschie, "A Review of E-Infinity Theory and the Mass Spectrum of High Energy Particle Physics," Chaos, Solitons & Fractals, Vol. 19, No. 1, 2004, pp. 209-236.doi:10.1016/S0960-0779(03)00278-9

11. R. L. Devaney, "An Introduction to Chaotic Dynamical Systems," Addison-Wesley, Redwood City, 1989.

12. M. S. El Naschie, O. E. Rossler and I. Prigogine, "Quantum Mechanics, Diffusion and Chaotic Fractals," Pergamon Press—Elsevier Publishing, Oxford, 1995.

13. L. B. Crowell, "Quantum Fluctutations of Space Time," World Scientific, Singapore City, 2005. doi:10.1142/5952

14. L. Glass and M. Mackey, "The Rhythms of Life," Princeton University Press, Princeton, 1988.

15. B. B. Mandelbrot, "The Fractal Geometry of Nature," Freeman, New York, 1983.

16. J. H. He, E. Goldfain, L. D. G. Sigalotti and A. Mejias, "Beyond the 2006 Physics Nobel Prize for COBE," China Culture and Science Publishing, Shanghai, 2006.

17. C. Beck, "Spation-Temporal Chaos and Vacuum Fluctuations of Quantized Fields," World Scientific, Singapore City, 2002. doi:10.1142/4853

18. Y. Baryshev and P. Terrikorpi, "Discovery of Cosmic Fractals," World Scientific, Singapore City, 2002. doi:10.1142/4896

19. J. Nicolis, G. Nicolis and C. Nicolis, "Non Linear Dynamics and the Two Slit Delayed Experiment," Chaos, Solitons & Fractals, Vol. 12, 2001, pp. 407-416. doi:10.1016/S0960-0779(00)00190-9

20. M. S. El Naschie, "Quantum Collapse of Wave Interference Pattern in the Two-Slit Experiment: A Set of Theoretical Resolution," Nonlinear Science Letter A, Vol. 2, No. 1, 2011, pp. 1-9.

21. R. Elwes, "Ultimate Logic," New Scientist, Vol. 211, No. 2183, 2011, pp. 30-33.doi:10.1016/S0262-4079(11)61838-1

22. J. Ambjorn, J. Jurkiewicz and R. Loll, "The Self-Organizing Universe," Scientific American, 2008, pp. 24-31.

23. S. Kranz and H. Park, "Geometric Integration Theory," Birkhauser, Boston, 2008.doi:10.1007/978-0-8176-4679-0

24. T. Jech, "Set Theory," Springer, Berlin, 2003.

25. A. Kanamori, "The Higher Infinite," Springer, Berlin, 2003.

26. A. Kechris, "Classical Descriptive Set Theory," Springer, New York, 1995.doi:10.1007/978-1-4612-4190-4

27. L. Graham and J. Kantor, "Naming Infinity," Harvard University Press, Cambridge, 2009.

28. L. M. Wapner, "The Pea and the Sun," A. K. Peters Ltd., Natick, 2005.

29. F. Morgan, "Geometric Measure Theory," Elsevier, Amsterdam, 2009.

30. M. S. El Naschie, "Quantum Entanglement as a Consequence of a Cantorian Micro Space-time Geometry," Journal of Quantum Information Science, Vol. 1, No. 2, 2011, pp. 50-53.doi:10.4236/jqis.2011.12007

31. G. Ord, M. S. El Naschie and J. H. He, Fractal SpaceTime and Non Commutative Geometry in High Energy Physics, Asian Academic Publishing Ltd., Hong Kong, Vol. 1, No. 1, 2011, pp. 1-46.

32. G. Ord, M. S. El Naschie and J. H. He, Fractal SpaceTime and Non Commutative Geometry in High Energy Physics, Asian Academic Publishing Ltd., Hong Kong, Vol. 2, No. 1, 2012, pp. 1-79.

33. L. Zadeh, "Fuzzy Logic and Approximate Reasoning," Synthese, Vol. 30, No. 3-4, 1975, pp. 407-428. doi:10.1007/BF00485052

34. L. Zadeh, "Fuzzy Sets," Information and Control, Vol. 8, 1965, pp. 338-353.doi:10.1016/S0019-9958(65)90241-X

35. K. Kosko, "Fuzzy Thinking," The New Science of Fuzzy Logic, Hyperion, New York, 1993.

36. M. S. El Naschie, "Fuzzy Knot interpretation of YangMills Instantons and Witten's 5 Brane Model," Chaos, Solitons & Fractals, Vol. 38, No. 5, 2008, pp. 1349-1354.doi:10.1016/j.chaos.2008.07.002

37. M. S. El Naschie, "From Experimental Quantum Optics to Quantum Gravity via a Fuzzy Kahler Manifold," Chaos, Solitons & Fractals, Vol. 25, No. 5, 2005, pp. 969-977.doi:10.1016/j.chaos.2005.02.028

38. M. S. El Naschie, "Fuzzy Dedochaedron Topology and E-Infinity Space-Time as a Model for Quantum Physics," Chaos, Solitons & Fractals, Vol. 30, No. 5, 2006, pp. 1025-1033.doi:10.1016/j.chaos.2006.05.088

39. N. M. Ahmed, "George Cantor: The Father of Set Theory," The Post Graduate Magazine, 2007, pp. 4-14.

40. M. S. El Naschie, "The Brain and E-Infinity", International Journal of Nonlinear Sciences and Numerical Simulation, Vol. 7, No. 2, 2006, pp. 128-131.

41. M.S.ElNaschie, "From Symmetry to Particle," Chaos, Solitons & Fractals, Vol. 32, No. 2, 2007, pp. 427-430. doi:10.1016/j.chaos.2006.09.016

42. M. S. El Naschie, "Kac-Moody Exceptional E12 from Simplectic Tiling," Chaos, Solitons & Fractals, Vol. 41, No. 4, 2009, pp. 1569-1571. doi:10.1016/j.chaos.2008.06.020

43. J. H. He, "Transfinite Physics," China Culture and Science Publishing, Shanghai, 2005.

44. M. S. El Naschie, "Knots and Exceptional Lie Groups as Building Blocks of High Energy Particle Physics," Chaos, Solitons & Fractals, Vol. 41, No. 4, 2009, pp. 1799-1803.doi:10.1016/j.chaos.2008.07.025

45. M. S. El Naschie, "Symmetry Group Prerequisite for E-Infinity in High Energy Physics," Chaos, Solitons & Fractals, Vol. 35, No. 1, 2008, pp. 202-211.doi:10.1016/j.chaos.2007.05.006

46. M. S. El Naschie, "Quantum Groups and Hamiltonian Sets on Nuclear Space-Time Cantorian Manifold," Chaos, Solitons & Fractals, Vol. 10, No. 7, 1999, pp. 1251-1256.doi:10.1016/S0960-0779(99)00009-0

47. M. S. El Naschie, "On a Class of Fuzzy Kahler-Like Manifold," Chaos, Solitons & Fractals, Vol. 26, No. 2, 2005, pp. 257-261. doi:10.1016/j.chaos.2004.12.024

48. M. S. El Naschie, "On a Class of General Theories for High Energy Particle Physics," Chaos, Solitons & Fractals, Vol. 14, No. 4, 2002, pp. 649-668. doi:10.1016/S0960-0779(02)00033-4

49. M. S. El Naschie, "On an Eleven Dimensional E-Infinity Fractal Space-Time," International Journal of Nonlinear Sciences and Numerical Simulation, Vol. 7, No. 4, 2006, pp. 407-409.doi:10.1515/IJNSNS.2006.7.4.407

50. M. S. El Naschie, "The Discrete Charm of Certain Eleven Dimensional Space-Time Theory," International Journal of Nonlinear Sciences and Numerical Simulation, Vol. 7, No. 4, 2006, pp. 477-481.

51. M. S. El Naschie, "On Fuzzy Kahler-Like Manifold Which Is Consistent with the Two-Slit Experiment," International Journal of Nonlinear Sciences and Numerical Simulation, Vol. 6, No. 2, 2005, pp. 8895-8898.

52. M. S. El Naschie, "The Symplectic Vacuum, Exotic Quasi Particles and Gravitational Instanton," Chaos, Solitons & Fractals, Vol. 22, No. 1, 2004, pp. 1-11.doi:10.1016/j.chaos.2004.01.015

53. J. H. He, "Non Linear Dynamics and the Nobel Prize in Physics," International Journal of Nonlinear Sciences and Numerical Simulation, Vol. 8, No. 1, 2007, pp. 1-4.doi:10.1515/IJNSNS.2007.8.1.1

54. L. Sigalotti and A. Mejias, "On El Naschie's Conjugate Complex Time, Fractal E-Infinity Space-Time and Faster than Light Particles," International Journal of Nonlinear Sciences and Numerical Simulation, Vol. 7, No. 4, 2006, pp. 467-472.doi:10.1515/IJNSNS.2006.7.4.467

55. M. Kaku, "Introduction to Superstrings and M-Theory," Springer, New York, 1999.

56. S. Weinberg, "The Quantum Theory of Fields," Cambridge, Vol. II, 1996.

57. S. Weinberg, "The Quantum Theory of Fields," Cambridge, Vol. III, 2000.

58. M. S. El Naschie, "Quantum Golden Field Theory—Ten Theorems and Various CONJECTURES," Chaos, Solitons & Fractals, Vol. 36, No. 5, 2008, pp. 1121-1125.doi:10.1016/j.chaos.2007.09.023

59. M. S. El Naschie, "An Outline for a Quantum Golden Field Theory," Chaos, Solitons & Fractals, Vol. 37, No. 2, 2008, pp. 317-323. doi:10.1016/j.chaos.2007.09.092

60. M. S. El Naschie, "Asymptotic Freedom and Unification in a Golden Field Theory," Chaos, Solitons & Fractals, Vol. 36, No. 3, 2008, pp. 521-525. doi:10.1016/j.chaos.2007.09.004

61. M. S. El Naschie, "A Guide to the Mathematics of E-Infinity Cantorian Space-Time Theory," Chaos, Solitons & Fractals, Vol. 25, No. 5, 2005, pp. 995-964.doi:10.1016/j.chaos.2004.12.033

62. M. S. El Naschie, "Elementary Prerequisites for E-Infinity," Chaos, Solitons & Fractals, Vol. 30, No. 3, 2006, pp. 579-605. doi:10.1016/j.chaos.2006.03.030

63. M. S. El Naschie, "The Theory of Cantorian Space-Time and High Energy Particle Physics (An Informal Review)," Chaos, Solitons & Fractals, Vol. 41, No. 5, 2009, pp. 2635-2646.doi:10.1016/j.chaos.2008.09.059

64. K. Svozil, "Quantum Field Theory on Fractal SpaceTime," Journal of Physics A, Vol. 20, No. 12, 1987, pp. 3861-3875. doi:10.1088/0305-4470/20/12/033

65. M. S. El Naschie, "Transfinite Harmonization by Taking the Dissonance Out of the Quantum Field Symphony," Chaos, Solitons & Fractals, Vol. 36, No. 4, 2008, pp. 781-786.doi:10.1016/j.chaos.2007.09.018

66. M. S. El Naschie, "Extended Renormalization Group Analysis for Quantum Gravity and Newton's Gravitational Constant," Chaos, Solitons & Fractals, Vol. 35, No. 3, 2008, pp. 425-431. doi:10.1016/j.chaos.2007.07.059

67. M. S. El Naschie, "Exact Non-Perturbative-Derivation of Gravity's G_4 Fine Structure Constant, the Mass of the Higgs and Elementary Black Holes," Chaos, Solitons & Fractals, Vol. 37, No. 2, 2008, pp. 346-359. doi:10.1016/j.chaos.2007.10.021

68. M. S. El Naschie, "Quantum E-Infinity Field Theoretical Gravitational Constant," International Journal of Nonlinear Sciences and Numerical Simulation, Vol. 8, No. 3, 2007, pp. 496-474.

69. M. S. El Naschie, "Towards a Quantum Golden Field theory," International Journal of Nonlinear Sciences and Numerical Simulation, Vol. 8, No. 4, 2007, pp. 477-482.doi:10.1515/IJNSNS.2007.8.4.477

Chapter 5

THE QUALITY MANAGEMENT OF THE R&D IN HIGH ENERGY PHYSICS DETECTOR

Xuemin Zhu[1] and Sen Qian[1]

[1]The Institute of High Energy Physics, Chinese Academy of Sciences, China

INTRODUCTION

Particle physics, also recognized as high energy physics, is a basic subject focusing on the research of the elementary elements of materials and their mutual actions. One distinguished characteristic of particle physics study is that the experimental equipments involved are always huge and special ones. Therefore, big science projects, including the R&D of large detectors, are usually required in high energy physics experiments. Those projects are complicated systematic engineering, involving many front and technology fields. It is impossible for a single institute to finish those large projects by its own. Cooperation among different institutes or organizations is necessary for big science projects in particle physics, especially international ones.

The Institute of High Energy Physics [1]

The Institute of High Energy Physics (IHEP) is the biggest and comprehensive fundamental research center in Chinese Academy of Science. The major research fields of IHEP are particle physics, accelerator physics and technologies, radiation technologies and application, Particle physics experiments and Accelerator physics and technology are two of the leading research areas. The main research facilities at IHEP include Beijing Electron Positron Collider (BEPC) and Beijing Spectrometer (BES), DayaBay Neutrino Experiment, Chinese Spallation Neutron Source, etc. IHEP has extensive cooperation with all high energy physics laboratories and participates in many important particle physics experiments in the world.

The Beijing Electron Positron Collider [2] and the Beijing Spectrometer [3]

The Beijing Electron Positron Collider (BEPC) consists of the injector, the storage ring, the transportation line, the Beijing Spectrometer (BES), the Beijing Synchrotron Radiation Facility (BSRF) and the computer center. Beijing Spectrometer (BES) is a general purpose magnetic spectrometer in the South IP of the storage ring. The general layout of the BEPC is shown in Fig.1.

Figure 1: The airscape of the BEPC.

BEPC started construction in 1984 and the first electron-proton collider was produced in Oct. 1988. BEPCII was installed in 2003 and finished five years later in 2009. IHEP establishes comprehensive and long-term cooperation with high energy laboratories and universities all over the world, especially in USA, Japan and Europe. With the international cooperation, IHEP have gained huge success in 30 years. For example, IHEP took part in the research of CMS and ATLAS detectors of Large Hadron Collider (LHC), which is the world's largest, highest-energy particle accelerator and the collider at the beginning of 21 centuries, built by CERN [4]. BESIII is also organized by IHEP and participated by 51 institutions and universities around the world, 34 from Asian, 12 from Europe and 5 from USA.

The DAYA Bay Neutrino Experiment [5]

The Daya Bay Neutrino Experiment is a neutrino-oscillation experiment designed to measure the mixing angle q13 using anti-neutrinos produced by the reactors of the Daya Bay Nuclear Power Plant (NPP) and the Ling Ao NPP.

The Daya Bay Neutrino Experiment is a major international joint research program, mainly organized by China working closely with researchers from other countries. In terms of both money and people, it is among the largest scientific collaborations between US and China. More than 200 scientists from China, include Hong Kong and Taiwan, the US, Russia, the Czech Republic are involved in the Daya Bay experiment. During the cooperation, China is in change of the laboratory construction, R&D of Anti-neutrino detector (AD), Gd-loaded Liquid Scintillator, Muon Veto Detector, Readout Electronic and Data acquisition system (DAQ) etc. While America is in responsible of the construction of water Cherenkov detector and so on.

Scientists from the Chinese Academy of Sciences (CAS) and the U.S.-based Brookhaven National Laboratory and the Lawrence Berkeley National Laboratory will participate in the underground experiment. An international funding commission comes into existence in the funding agency to discuss fee issues and instruct the experiment process and fee management through experimental supervision organization. The project management of the Daya Bay Neutrino Experiment adopts the advanced and mature modern management idea used for managing large international joint project and big science experimental research project. An international cooperation group is built and management rules are made. Besides, a cooperation group commission is founded, during which executive board and spokesperson is elected for overall supervision of the whole project. The Daya Bay Neutrino Experiment is initiated in 2007 and finished in 2012.

Chinese Spallation Neutron Source [6]

Chinese Spallation Neutron Source (CSNS) is designed to build a device with the power of proton beam reaching up to 100 kW effective and the flux of pulsed neutrons coming out top in the world, along with other three spallation neutron sources built in America, Japan and British. CSNS is also a large cooperative project, supported by Chinese Academy of Sciences and Guangdong government. The normal operation for uses is foreseen in 2018. IHEP is the main construction institution in the project with the Institute of Physics Chinese Academy of Sciences as the co-operation unit. The construction team bring together three generations of outstanding scientific and technical researchers in China. An international CSNS neutron technology advisory committee is set up for reviewing the key experimental work. The experts of the advisory committee are from well-known laboratories in America, Japan, Germany, Australia and other countries.

INTRODUCTION OF QUALITY MANAGEMENT OF SCIENTIFIC PROJECTS IN IHEP

During the process of big science project and research, IHEP has significant advantages in accelerator physics and technology, human resources, international cooperation and academic exchange. IHEP owns mature model and advanced experience in the quality management of scientific projects.

Project Management System

Before 2011, the project manager is responsible for the big science project management in IHEP. International cooperation group is formed and fees are under the sponsors' supervision and review. There is a perfect project management system, though without quality management system meeting international standards.

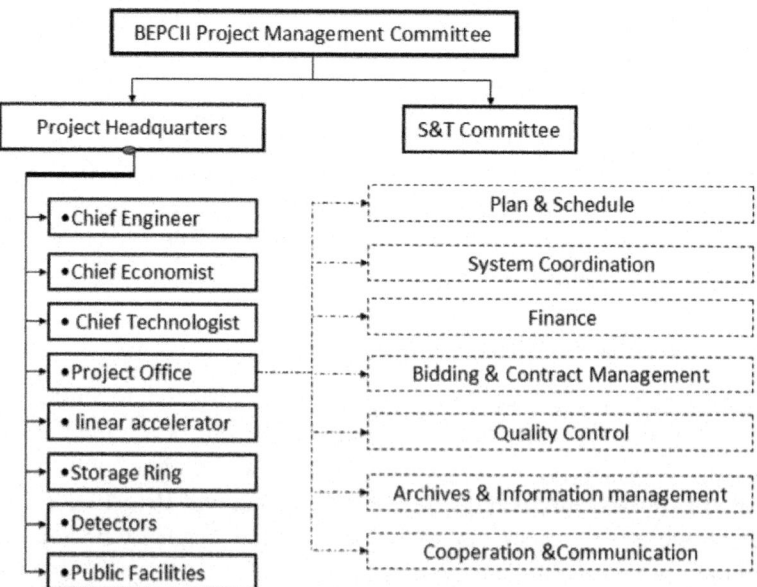

Figure 2: The Organization Chart of the BEPCII Project Management.

In the project management system shown in Fig.2, special-purpose management mechanism such as fund, purchasing, quality, safety and archive is established, with clear responsibilities and authorities. Besides, the internal communication mechanism and interface management mechanism are also set up. CPM Plan is adopted for fund and schedule management. To ensure the quality of the project, during the design and development process, experts are always invited for evaluation. And an international council committee is asked for review in terms of major international cooperation projects.

In fact, the requirements of the project management system have already displayed in the ISO 9001 quality management system. Though without a systematic quality manual and standards and lack of resource, purchasing and archive management. In the project management system, quality management is more focused on the management of various test guidelines and processing of key parts (including outsourced progress)

Quality Management Systems

The BEPCII project headquarters has placed great important on the quality management and published "BEPCII project management file" in 2002. In the file, responsibilities and rights of personnel, fund management, file number, document signing and alteration, early stages management, bidding and purchasing are described in detail.

At the beginning of 2005, during the construction of BEPCII project, the headquarters built a quality management system according to GB/T19001-2000(idt ISO9001:2000). Although the system doesn't get a national certification, it is completely in accordance with standards of quality management system requirements and it has played a very good effect. In 2009, BEPCII completed the construction task successfully by time, with high quality and budget under control.

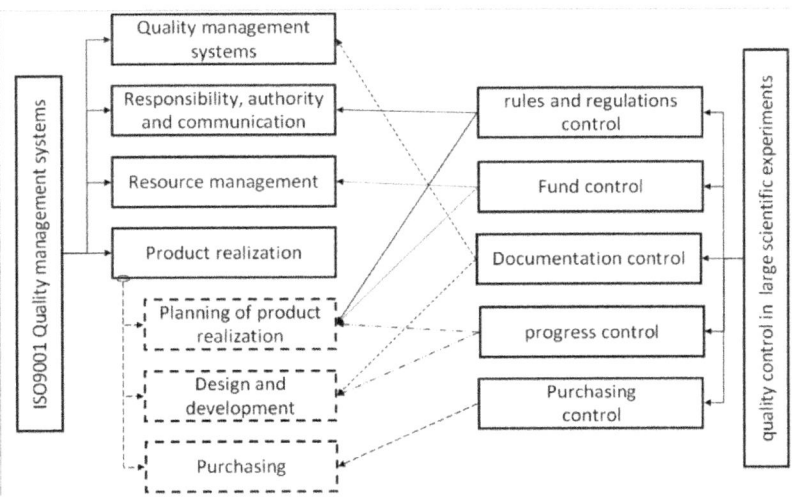

Figure 3: The Relationship between the quality control system in large scientific experiments and ISO9001 Quality management systems.

As described in Fig.3, The quality control in large scientific experiments corresponds with ISO9001 Quality management systems, which is classified according to the production. While, the quality control in large scientific experiments is classified according to the type of different work. The ISO9001 Quality management system is widely adopted by corporations all over the word and it's more normative.

In 2011, IHEP passed a national quality management certification system: GB/T19001-2008(idt ISO9001 : 2008). After two years' development of quality management system from its very beginning to being passed, it has confirmed that IHEP has the ability to produce scientific production meeting requirements.

The set up of quality management system makes the project management procedure standard, and promotes the overall management level in IHEP. The clients' needs are fully met and the quality management of IHEP joined the line of international standard management. The role played by quality management in the scientific research, especially in the big science project, is invaluable and imponderable.

QUALITY MANAGEMENT IN R&D OF BESIII DETECTOR

The project of BESIII detector began its research and development, according to the scientific project management system and quality management system, like other big science projects.

Mechanism Management

BESIII detector R&D is part of BEPCII project. So the quality management of the detector research is responsible by the project director. As a whole, BESIII carries out the management system of BEPCII project headquarters strictly and makes some special mechanism to form a mechanism with a clear hierarchy. Quality technician are employed in the project.

BESIII detector R&D project has outlined the responsibilities and rights of each person in charge with an appropriate staffing in the organization. The communication methods of the total and sub system and record control requirements are defined.

The director in charge of sub system is responsible for the implementation of the BESIII research plan, management, arrangement of related resources and coordination with scientific and technical issues. Each division leading person is specifically responsible for the respective task implementation plan. Members in the project cooperate with each other closely at reaching difficult goals. The whole project has the characteristic of unified task, defined responsibilities, reasonable arrangement and integrated resources.

The high energy physics experiment is a complex project, and the communication in different study cells seems more important. The Task Control Form is widely used in study works, and the forms are preserved and archived as records of the system.

Subject				
Send to				
From		Date		
Serial No.		Pages		
Attached				
C.C.				
Content				
Jointly Sign				
validation	Prepared by	Checked by	Examined by	Approved by
Signature				
Date				

Figure 4: The task control form used by different teams.

Researchers in the project communicate with each other in time and have a regular meeting each or twice a week, to make sure the project is under schedule control and discuss some technical problems. Meeting minutes are kept as a reference. Sub-system will report the progress of the project and accept an inspection and evaluation regularly.

Fund Management

Fund management is important for the whole management of scientific project. Appropriate fund use a basis for carrying out any high energy physics experiment smoothly. As for the R&D of BESIII detector, the experiment design and development planning will affect the rationality of the budget and fund use directly. They are also the important contents in the requirements of the quality management system

Funds come from Chinese Academy Sciences (CAS) allocation and self-provided funds in the BESIII project. At the end of the year, expenses are counted and reported to CAS and the project will receive examination and evaluation.

Control of Documents and Records

Control of documents and records is critical whether for scientific project management or for quality management. For high energy physic experiments, large and complicated equipments are usually involved. During the project design and scheme phase, rules for documents and records reserved need to be made clearly and principles for numbering and signing the documents and records need to be described specifically.

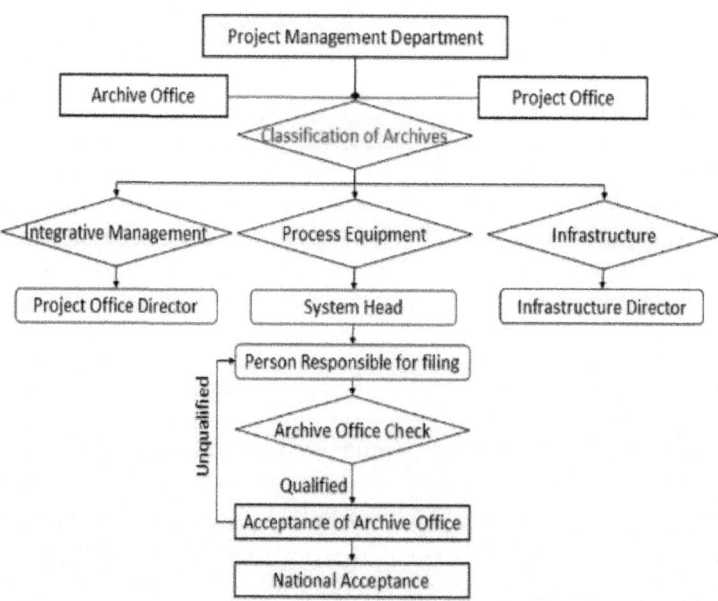

Figure 5: The workflow of archive management.

BEPCII project builds up a special mechanism of file control. Because BESIII is part of the whole project, the rules of document management are in accordance with the requirements of BEPCII.*Documents and records* need to be signed according to the regulations, in accordance with the whole project and effective as well. *Documents and records* need to be preserved and archived on a regular basis.

There are several characteristics in archive management work, especially for the high energy experiments. Firstly, this work must be arranged by the project management department at the beginning of the project. Secondly, the document and records which need to be preserved must be clearly described and the responsibility should be defined at the first time. Thirdly, the archive office, the project office and each member working for the project should cooperate to get the work done quickly and perfectly.

All the quality documents of the whole process of each single detector, from design, research, test, and acceptance are preserved according to the regulations. Technical specifications, interface of different tasks, diagrams, test reports are archived as written documents. Regular meeting minutes are kept also as archives. Those *Documents and records* can be used to track and follow the quality of the product in the whole process.

The running cycle of big science project, just like its construction cycle, is as long as to last more than ten years. Therefore, control of documents and records is very essential for the running and maintenance of the big science project, as an important prop and support.

Schedule Management

BESIII project has followed Critical Path Method (CPM) to control the schedule of the whole project. The plan in the CPM is in detail and convenient for check. It is easy for revision according to the actual process and make sure it is updated in time within the system.

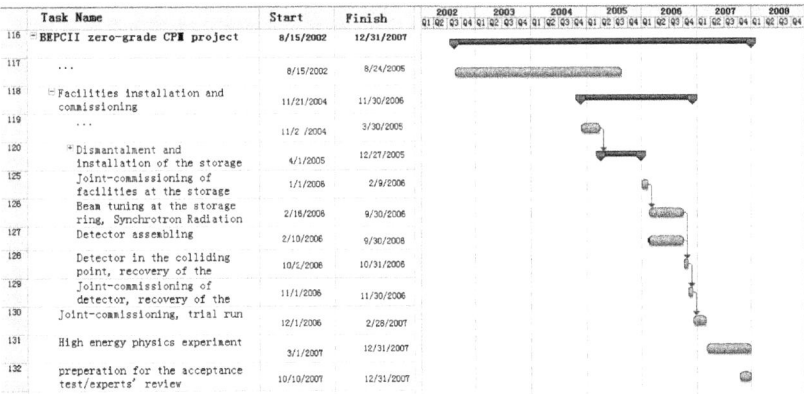

Figure 6: BEPCII zero-grade CPM project (partly,2002).

In order to give a better control of the schedule, CPM is classified. Any sub-system could make its own play and updates in time following the step of the total plan. Therefore inter communication plays an important role in the schedule management. In a word, CPM is a further refinement of the time arrangement of the project design report and makes the management of the project construction effective.

The CPM project is highly in accord with the practical progress, and, BEPCII zero-grade CPM project is modified frequently. The BEPCII project was finished in 2008 and was finally checked in 2009.

Purchasing Management

The R&D of large detectors is involved with bulk purchase. In the BESIII project, purchasing management rules are made according to the relevant laws and regulations on acquisition. Purchasing and approval process are defined clearly. Bidding is strictly adopted in the project to save research money. An appropriate regulation in the purchasing process is a guarantee for carrying out the project under the budget..

Abroad purchase has a long life cycle, heads of procurement need to do significant preparatory work in advance, and the heads should be quite familiar with the procurement procedures in order to complete the purchase in time. The purchasing department published the flowchart to facilitate the work.

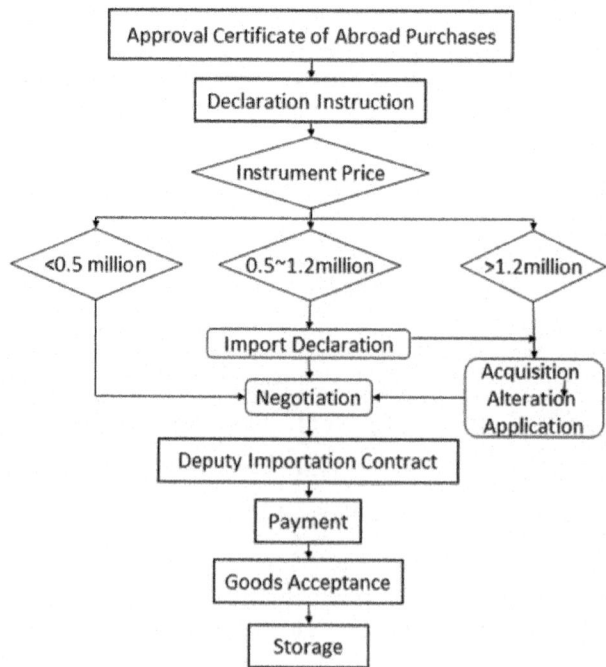

Figure 7: Abroad purchase flowchart.

BESIII-MUC QUALITY MANAGEMENT IN R&D OF BESI-II-MUC DETECTOR

Introduction of BESIII Detector and MUC Detector

The Beijing Spectrometer (BESIII) is designed to measure the properties of the particles produced in the collisions of electrons and positrons at BEPCII. The physics goal of the BESIII experiment is to conduct high statistics and highly precise studies on a number of physics topics in this energy region, including light hadron spectroscopy, charmonium spectra, charm meson decay properties, QCD, tau physics, rare decays, search for glueballs and other non-pure quark states [3].

The BESIII detector will consist of a 1 T superconducting solenoid magnet, a high precision main drift chamber (MDC), Time-Of-Flight counters (TOF), a CsI crystal Electromagnetic Calorimeter (EMC) and a muon identifier chamber (MUC) that is integrated in the iron magnetic field return yoke [7]. The muon identifier is the outer most subsystem of the BESIII detector [8], which is constructed by resistive plate chambers (RPCs, shown in Fig.8.a). 962 RPC are used in the whole MUC detector, which consists of 136 RPC superlayer modules (SM, shown in Fig.8.b). And the Fig.8.c shows the status of the MUC detector when it was finished it's barrel part assemblage. The Fig.8.d shown the designed construct of the BESIII MUC detector with the endcap and barrel parts.

(a)

Figure 8: a). The RPC moduls, (b). The Suprlayer Modul, (c). The overview the barrel part of the MUC detector after it's assemblage, (d) The construct of the BESIII MUC detector.

Quality Management in the R&D of BESIII-MUC Detector

The whole process of R&D of MUC detector include the design of the basic unit RPC, properties investigation, bulk production, SM design; design of MUC detector, installation debugging, running and maintenance.

Throughout the research process, the project director managed the project scientifically and effectively, with each research aspect considered carefully, comprehensively and deeply, and made some achievements. From the pre-research in 2003 to the formal data collection in 2009, more than 30 papers have been published by the research group of MUC, covering the whole research process.

7 papers have been published in *NIMA*, as follows:

1. A new surface treatment for the prototype RPCs[9],
2. Cosmic ray test results on resistive plate chamber for the BESIII experiments [10]
3. The Design and Mass Production on RPC for the BESIII Experiment [11]
4. A monitor for the composition of the gas mixture of BESIII muon chambers [12]
5   .First results of the RPC commissioning at BESIII [13]
6. The BESIII Muon Identification System [14]
7. An underground cosmic-ray detector made of RPC [15]

8 papers have been published in *Chinese Physics C,*as follows:

1. Cosmic Ray Test Station for BESIII RPC [16]
2. Research and Development of Large Area Resistive Plate Chamber [17]
3. A Study of RPC Gas Composition using Daya Bay RPCs [18]
4. Quality control and database on RPC for the BESIII experiment [19]
5. Test of BESIII RPC in the avalanche mode [20]
6. Performance Study of RPC Prototypes for the BESIII Muon Detector [21]
7. Study of the RPC-Gd as thremal neutron detector [22]

Table 1: The analysis of the manuscripts published by MUC group

	Design	Performance Test	Mass Production	Research Work	Application
RPC	1 b	2 a c e f	3 d	4	7 g
SM			6		
MUC			6	5	

As shown in Table 1, it is not difficult to come into conclusion that the whole R&D of MUC detector applied scientific project quality management, which promotes the research work. In the phase of initial RPC research, the key point is on the study of the detector's performance test. It is the phase for building a standard quality management. After the acceptance of RPC and project review, mass production and SM reassembling come into being. In this phase, scientific quality control and management play a key role.

A perfect quality tracking system is established in each session, from the production and test of RPC, assembling and test of modules, to the installation and debugging of MUC detector, to ensure the supervision of the performance of detector is plausible.

Especially for the mass production of RPC and SM, before research and test, a database is built for storage related data and affording date support for quality control and final running & maintenance.

Summary

All the requirements such as verification, validation, monitoring, measurement, inspection and test activities specific to the detector are described in the design report of the detector in detail. The report plays the same role as in making a particular quality control plan.

Table 2: Quality management/control factor distribution of MUC detector

□ ■	Design	Proto-type	Mass pro-duction	Assem-blage	Debug	Running
critical charac-teristic	■	□				
major charac-teristic	■	■	□			
critical pro-cess		■	■			

article inspection	■	■	■			■
quality improvement	■	■		■		
effectiveness			■	■		
traceability			■	■	■	■
preventive action		□	□		■	■
corrective action					■	
quality plan			□	■		

The design report of the detector divide the R&D process into several phases, including concept design, project design, sample trail-manufacture, product research and production, test, installation and debugging. In each phase, review and identification is defined. For important phases, such as aging test, assembly test and system test, detailed guidelines and instructor are written. As shown in Table 2, during the outsourcing process, key parts are defined, and acceptance rules are also clearly described. Control point is set up and design files are carried out strictly to ensure the product quality. More detailed could be found in table 2 for summary.

SIGNIFICANCE OF SCIENTIFIC QUALITY MANAGEMENT IN RESEARCH

Promote Scientific Projects

We could come into conclusion that scientific quality management can promote scientific projects to proceed successfully, in the following ways:

The schedule of the project could be arranged and controlled well, especially the adoption of CPM, which could provide a time map for the whole project. Throughout the four years' successful implementation and of BEPCII project, CPM plays an important role in the project acceptance in due. CPM was adjusted in time according to the project status, thus effective management and restriction was formed for all the related sub systems in the project.

The project has been implemented within the budget and cost was controlled. Purchasing procedures and approval process were strictly described, which played a role for the fair use of the fund.

Documents and records were kept in detail, as reference in the project to find the source of old problems and avoid new problems. Especially for those big scientific projects which will last more than ten years, files about interface management and quality management and various records are significant for the running and maintenance in the following work. They also act as important reference for the future project construction in high energy physics.

Promote Scientific Research

Scientific quality management could promote scientific research effectively. At the same time, as the development of scientific research, cooperation among researchers will be increased. It is good for the communication and exchange in the area of quality management and promotes the refining of the quality management system thus.

Experiences in big scientific project are good for the growth of young researchers. With participation in the R&D of big science equipment under quality management system, researchers will learn how to organize and manage scientific programs or projects in future.

In an ongoing scientific project managed launched by IHEP, researchers are from participants in BESIII or DayaBay. Although it is non-international, at the beginning the project is managed as required in strict quality management, just like that in big science project. As the development of the project, communication and cooperation among other institutions both at home and abroad have increased. To coordinate the partnership among different organizations and unites, cooperation group is formed. As shown below, a formality management system and strictness organization is built, which lays a solid foundation for the sustainable development of cooperation group and joint research work in future, whose Organization Chart shown in fig.9 for example.

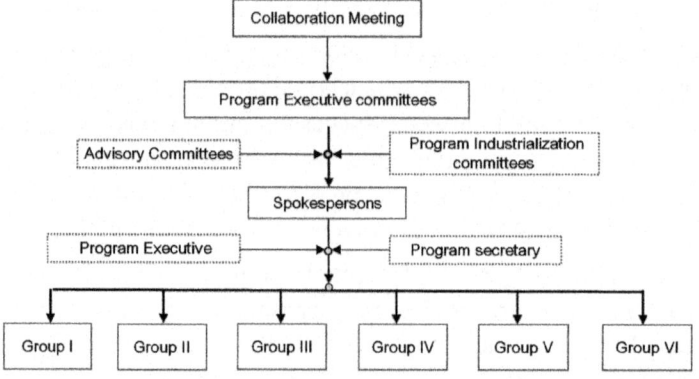

Figure 9: The organization of the BEPCII.

CONCLUSION

Quality management plays a significant role both in project management and in the scientific research. With a scientific and comprehensive quality management system, big science project will be duly executed. The level of scientific projects will be greatly improved by the application and popularization of national and international quality standards.

ACKNOWLEDGEMENTS

In the process of writing this manuscript, we have received much understanding and support from many departments in our institute. We are particularly grateful to those staffs in the IHEP offices, archives, the purchasing department and the project teams for their strongly supported. Special thanks go to Prof. Zhao Jingwei for this support and encouragement to finish this work.

REFERENCES

1. The Institute of High Energy Physics. http://english.ihep.cas.cn/au/bi/

2. The BEPCII Project. http://english.ihep.cas.cn/rs/fs/bepc/index.html

3. The Beijing Spectrometer. http://bes3.ihep.ac.cn/orga/institute.htm

4. European Organization for Nuclear Research. http://public.web.cern.ch/public/

5. The Bay Reactor Neutrino Experiment. http://dayabay.ihep.ac.cn/twiki/bin/view/Public/WebHome

6. China Spallation Neutron Source (CSNS). http://csns.ihep.ac.cn/english/index.htm

7. Z. Tianchi, Design and construction of the BESIII detector Nuclear. Nuclear Instruments and Methods in Physics Research A 614 2010 345 399

8. Y. Boxiang, The construction of the BESIII experiment. Nuclear Instruments and Methods in Physics Research A 598 2009 7 11

9. Z. Jiawen, A new surface treatment for the prototype RPCs. Nuclear Instruments and Methods in Physics Research A 540 2005 2005 102 112

10. H. Jifeng, Cosmic ray test results on resistive plate chamber for the BESIII experiments Cosmic ray test results on resistive plate chamber for the BESIII experiments.

11. Z. Jiawen, The Design and Mass Production on RPC for the BESIII Experiment. Nuclear Instruments and Methods in Physics Research A 580 2007 1250 1256

12. Q. Sen, A monitor for the composition of the gas mixture of BESIII muon chambers. Nuclear Instruments and Methods in Physics Research A 595 2008 520 525

13. X. Yuguang, First results of the RPC commissioning at BESIII Nuclear Instruments and Methods in Physics Research A 595 2008 520 525

14. Q. Sen, The BESIII Muon Identification System. Nuclear Instruments and Methods in Physics Research A 614 2010 196 205

15. Z. Qingmin, An underground cosmic-ray detector made of RPC Nuclear Instruments and Methods in Physics Research A 583 2007 278 284

16. L. Qian, Cosmic Ray Test Station for BESIII RPC. China Physics C (High Energy And Unclear Physics) 2006 30 4

17. Z. Jiawen, Research and Development of Large Area Resistive Plate Chamber. China Physics C (High Energy And Unclear Physics) 2003 27 7

18. H. Malie, Study of RPC gas composition using Daya Bay RPCs. China Physics C (High Energy And Unclear Physics) 2010 34 8

19. H. Jifeng, Quality control and database on RPC for the BESIII experiment. China Physics C (High Energy And Unclear Physics) 2008 32 3

20. H. Jifeng, Test of BESIII RPC in the avalanche mode. China Physics C (High Energy And Unclear Physics) 2008 32 5

21. X. Yuguang, Performance Study of RPC Prototypes for the BESIII Muon Detector. China Physics C (High Energy And Unclear Physics) 2008 31 1

22. Q. Sen, Study of the RPC-Gd as thremal neutron detector. China Physics C (High Energy And Unclear Physics) 2009 33 8

Chapter 6

GRID COMPUTING IN HIGH ENERGY PHYSICS EXPERIMENTS

Dagmar Adamová[1] and Pablo Saiz[2]

[1]Nuclear Physics Institute, Řež near Prague ˇ Czech Republic

[2]CERN Switzerland

INTRODUCTION

The High Energy Physics (HEP) [1] – often called Particle Physics – is one of the research areas where the accomplishment of scientific results is inconceivable without the infrastructure for distributed computing, the Computing Grid. The HEP is a branch of Physics that studies properties of elementary subatomic constituents of matter. It goes beyond protons and neutrons to study particles which existed only a fraction of a second after the Big Bang and quarks and gluons in the so-called Quark Gluon Plasma (QGP) [2]. These studies are based on experiments with particles colliding at very high energies, at speeds almost equal to the speed of light.

The world's leading Particle Physics research laboratory is CERN [3], the European Center for Nuclear and Particle Physics near Geneva, Switzerland. The CERN latest particle accelerator (see Fig. 1), the Large Hadron Collider (LHC) [4], installed in a 27 km long tunnel located about 100 m underground and crossing the Swiss - French border, uses counter-rotating beams of protons or lead ions (Pb) to collide at 4 points, inside large particle detectors: ALICE [5], ATLAS [6], CMS [7] and LHCb [8]. There are also another two smaller experiments, TOTEM [9] and LHCf [10]. These are much smaller in size and are designed to focus on "forward particles". These are particles that just brush past each other as the beams collide, rather than meeting head-on.

Figure 1: LHC@CERN.

The energy of the protons is currently 3.5 TeV (1 Tera eV= 1 million MeV) and that of the Pb ions is 1.38 TeV, so the collision energies are 7 TeV for the protons and 2.76 TeV for the Pb ions. The phrase often used to summarize the mission of the LHC is, that with the LHC we are going back in time very close to the Big Bang, as close as about 10−10 seconds. In terms of length it represents about 10−16 cm (compared to the dimensions of the Universe of about 1028 cm). At this scale, the matter existed in a form of a "soup" made of the quarks and gluons, the Quark Gluon Plasma. The quarks are objects protons and neutrons are made of, so the LHC represents in a sense a huge extremely complicated microscope enabling the study of the most basic elements of matter.

There are several major questions which scientists hope to get answered with the help of the LHC.

- What is the origin of mass, why do elementary particles have some weight? And why do some particles have no mass at all? At present, we have no established answers to these questions. The theory offering a widely accepted explanation, the Standard Model [11], assumes the existence of a so-called Higgs boson, a key particle undiscovered so far, although it was first hypothesized in 1964. One of the basic tasks of the LHC is to bring an established statement concerning the existence of the Higgs boson.

- Where did all the anti-matter disappear? We are living in the World where everything is made of matter. We suppose that at the start of the Universe, equal amounts of matter and antimatter were produced in the Big Bang. But during the early stages of the Universe, an un-known deviation or in-equilibrium must have happened, resulting in the fact that in our world today hardly any antimatter is left.

- What are the basic properties of the Quark-Gluon Plasma, the state of the matter existing for a tiny period of time after the Big Bang? Originally, we thought it would behave like a plasma, but the latest scientific results including those delivered by the LHC suggest that it behaves like a perfect liquid [2], which is somewhat surprising for us.

• What is the universe made of? At the moment, the particles that we understand create only 4 % of the universe. The rest is believed to be made out of dark matter and dark energy. The LHC experiments will look for supersymmetric particles, which would confirm a likely hypothesis for the creation of dark matter.

From the experiments analyzing the data from the LHC collisions, ATLAS and CMS are the largest. They were nominally designed to look for the Higgs boson but in fact these are general purpose detectors for the study of all kinds of Physics phenomena at the LHC energy range. The ALICE detector is a dedicated heavy ions detector to study the properties of the Quark Gluon Plasma formed in the collisions of lead ions at the LHC energies. The LHCb is much smaller detector and its mission is to study the asymmetry between matter and antimatter. Although all these experiments are designed for Particle Physics research, the scientific programs they follow actually cross a border between Particle Physics, Astrophysics and Cosmology.

Now, where does the Computing Grid show up in this scientific set-up? The LHC is the world's largest particle accelerator. The protons and lead ions are injected into the acceleratorin bunches, in counter-rotating beams. According to the original design proposal, there should be 2808 bunches per a beam. Each bunch of protons contains 1011 protons. the design beam energy is 7 TeV and the design luminosity is 10^{34} cm^{-2} s^{-1}. The bunch crossing rate is 40 MegaHz and the proton collisions rate $10^{7} - 10^{9}$ Hz.

However, the new phenomena looked for by the scientists appear at a rate of 10^{-5} Hz. So the physicists must analyze 10^{13} collision events/sec to have a chance to discover a New Physics phenomenon. At present, the machine has not yet reached the full number of bunches per beam and is operating at half of the originally proposed energy, but the luminosity is getting rapidly to the goal value. The LHC team has been increasing the number of bunches gradually reaching 1380 bunches/beam at the time of writing. The full beam energy will be reached in 2014, after one year of a technical stop to arrange for this increase. The machine has already beaten some world records which we will mention in section 5. Let us just mention the one concerning the stored energy: at the end of 2010, the energy stored in the accelerator ring was about 28 MegaJoules (MJ). At the target full intensity, this energy will reach about 130 MJ which is an equivalent of 80 kg of TNT.

In any case, the volume of data necessary to analyze to discover New Physics was already in the original proposal estimated to be about 15 PetaBytes (PB, 1PB=1 million GB) per data taking year. The number of the processor cores, CPUs, needed to process this amount of data was estimated to be about 200 thousands. And here, the concept of a distributed data management infrastructure comes into the scenario, because there is no single computing center within the LHC community/collaboration to offer such massive computing resources, even not CERN. Therefore in 2002, the concept of the Worldwide LHC Computing Grid (WLCG) [12] was launched to build a distributed computing infrastructure to provide the production and analysis environments for the LHC experiments.

In the present chapter, we give a short overview of the Grid computing for the experiments at the LHC and the basics of the mission of the WLCG. Since we are members of the ALICE collaboration, we will also describe some specific features of the ALICE distributed computing environment.

In section 2, we will describe the architecture of the WLCG, which consist of an agreed set of services and applications running on the Grid infrastructures provided by the WLCG partners. In section 3, we will mention some of the middleware services provided by the WLCG which are used for the data access, processing, transfer and storage. Although WLCG depends on the underlying Internet - computer and communications networks, it is the special kind of software, so-called middleware, that enables the user to access computers distributed over the network. It is called "middleware" because it sits between the operating systems of the computers and the Physics applications that solve particular problems. In section 4, the Computing model of the ALICE experiment will be briefly described. It provides guide lines for the implementation and deployment of the ALICE software and computing infrastructure over the resources within the ALICE Grid and includes planning/ estimates of the amount of needed computing resources. Section 5 will be devoted to the ALICE-specific Grid services and the ALICE Grid middleware AliEn. It is a set of tools and services which represents an implementation of the ALICE distributed computing environment integrated in the WLCG environment. In section 6, an overview will be given of the experience and performance of the WLCG project and also of the ALICE Grid project in particular during the real LHC data taking. The continuous operation of the LHC started in November 2009.

When the data started to flow from the detectors, the distributed data handling machinery was performing almost flawlessly as a result of many years of a gradual development, upgrades and stress-testing prior to the LHC startup. As a result of the astounding performance of WLCG, a significant number of people are doing analysis on the Grid, all the resources are being used up to the limits and the scientific papers are produced with an unprecedented speed within weeks after the data was recorded.

Section 7 contains a short summary and an outlook. This chapter is meant to be a short overview of the facts concerning the Grid computing for HEP experiments, in particular for the experiments at the CERN LHC. The first one and half a year of the LHC operations have shown that WLCG has built a true, well functioning distributed infrastructure and the LHC experiments have used it to rapidly deliver Physics results. The existing WLCG infrastructure has been and will be continuously developing into the future absorbing and giving rise to new technologies, like the advances in networking, storage systems, middleware services and inter-operability between Grids and Clouds.

WLCG

As mentioned in section 1, the LHC experiments are designed to search for rare events with the signal/noise ratio as low as 10^{-13}. This Physics requires a study of enormous number of pp and Pb-Pb collisions resulting in the production of data volumes of more than 10 PetaBytes per one data taking year. The original estimates elaborated when the LCG TDR [13] was put together were about ~ 15 PetaBytes (PB) of new data each year which translates into ~ 200 thousands of CPUs/processor cores and 45 PB of disk storage to keep the raw, processed and simulated data.

Nowadays, 200 thousands cores does not sound like much and one can find them in large computer centers. 50 PB of a disk storage is however not that common. In any case, at the time the LHC Computing Grid was launched there was no single site within the LHC community able to provide such computing power. So, the task of processing the LHC data has been a distributed computing problem right from the start.

WLCG Mission

The Worldwide LHC Computing Grid (WLCG) project [13] was launched in 2002 to provide a global computing infrastructure to store, distribute and process the data annually generated by the LHC. It integrates thousands of computers and storage systems in hundreds of data centers worldwide, see Figure 2. CERN itself provides only about 20% of the resources needed to manage the LHC data. The rest is provided by the member states' national computing centers and research network structures supported by national funding agencies. The aim of the project is the "collaborative resource sharing" between all the scientists participating in the LHC experiments, which is the basic concept of a Computing Grid as defined in [14]. The infrastructure is managed and operated by a collaboration between the experiments and the participating computer centers to make use of the resources no matter where they are located.

The collaboration is truly worldwide: it involves 35 countries on 5 continents and represents 49 funding agencies having signed the WLCG Memorandum of Understanding on Computing (WLCG MoUC) [15]. The distributed character has also a sociological aspect: even if the contribution of the countries depends on their capabilities, a member of any institution involved can access and analyze the LHC data from his/her institute.

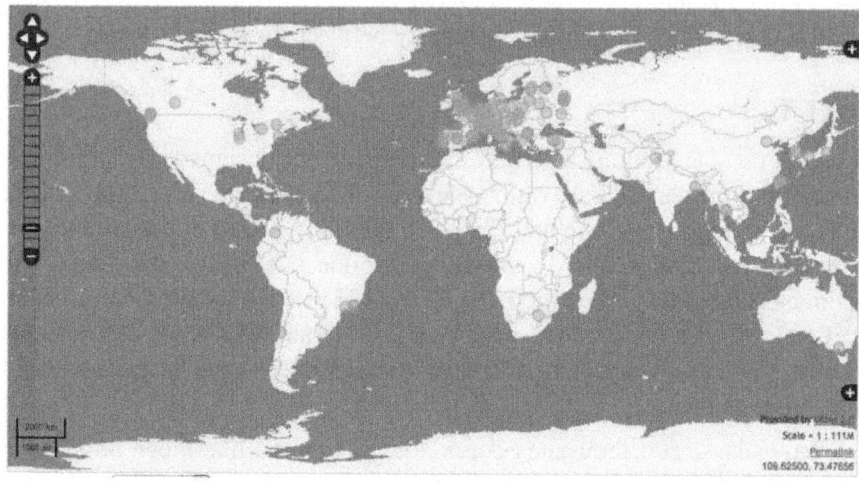

Figure 2: Distribution of WLCG computing centers.

Currently, the WLCG integrates over 140 computing sites, more than 250 thousands CPU cores and over 150 PB of disk storage. It is now the world's largest computing grid: the WLCG operates resources provided by other collaborating grid projects: either the two main global grids, EGI [16] and OSG [17], or by several regional or national grids.

Hierarchical (Tier) Structure, the Roles of Different Tier-Sites

The WLCG has a hierarchical structure based on the recommendations of the MONARC project [18], see Figure 3. The individual participating sites are classified according to their resources and level of provided services into several categories called Tiers. There is one Tier-0 site which is CERN, then 11 Tier-1 centers, which are large computing centers with thousands of CPUs, PBs of disk storage, tape storage systems and 24/7 Grid support service (Canada:

TRIUMF, France: IN2P3, Germany: KIT/FZK, Italy: INFN, Netherlands: NIKHEF/SARA, Nordic countries: Nordic Datagrid Facility (NDGF), Spain: Port d'Informació Científica (PIC), Taipei: ASGC, United Kingdom: GridPP, USA: Fermilab-CMS and BNL ATLAS). Then there are currently about 140 Tier-2 sites covering most of the globe. The system also recognizes Tier-3 centers, which are small local computing clusters at universities or research institutes.

The raw data recorded by the LHC experiments (raw data) is shipped at first to the CERN Computing Center (CC) through dedicated links. CERN Tier-0 accepts data at average of 2.6 GBytes(GB)/s with peaks up to 11 GB/s. At CERN, the data is archived in the CERN tape system CASTOR [19] and goes through the first level of processing - the first pass of reconstruction. The raw data is also replicated to the Tier-1 centers, so there are always 2 copies of the raw data files. CERN serves data at average of 7 GB/s with peaks up to 25 GB/s [20]. The Tier-0 writes on average 2 PB of data per month to tape in pp running, and double that in the 1 month of Pb-Pb collisions, (cf. Figures 4,5). At Tier-1 centers, the raw data replicas are permanently stored as mentioned before and several passes of the data re-processing are performed. This multiple-stage data re-processing is performed using methods to detectinteresting events through the processing algorithms, as well as improvements in detector calibration, which are in continuous evolution and development. Also, the scheduled analysis productions as well as some of the end user analysis jobs are performed at Tier-1s.

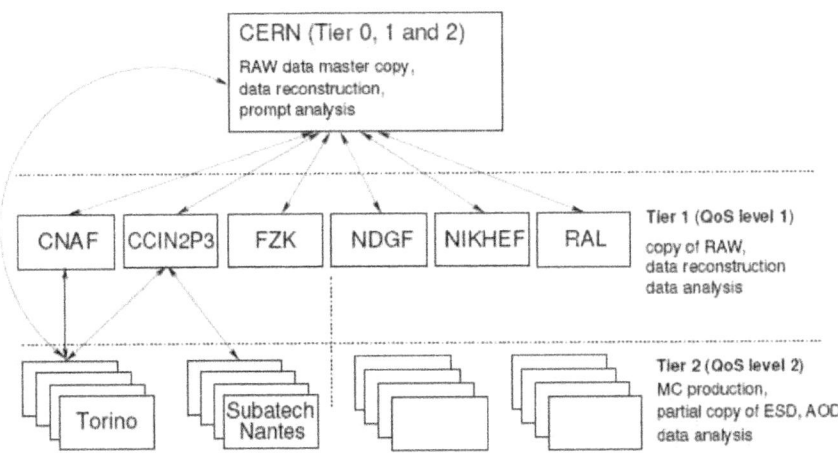

Figure 3: Schema of the hierarchical Tier-like structure of WLCG.

Tier-2 centers (more than 130 in the WLCG, integrated within 68 Tier-2 federations) are supposed to process simulation (Monte Carlo simulations of the collision events in the LHC detectors) and end-user analysis jobs. The load of simulations needed to correctly interpret the LHC data is quite sizeable, close to the raw data volume. The number of end users regularly using the WLCG infrastructure to perform analysis is larger than expected in the beginning of the LCG project, it varies from about 250 to 800 people depending on the experiment. This is certainly also a result of the experiments' effort to hide the complexity of the Grid from the users and make the usage as simple as possible. Tier-2 sites deliver more than 50% of the total CPU power within the WLCG, see Figure 6.

Network

The sustainable operation of the data storing and processing machinery would not be possible without a reliable network infrastructure. In the beginning of the WLCG project there were worries that the infrastructure would not be able to transfer the data fast enough. The original estimates of the needed rate were about 1.3 GB/s from CERN to external Tiers. After the years spent with building the backbone of the WLCG network, CERN is able to reach rates about 5 GB/s to Tier-1s, see Figure 7. The WLCG networking relies on the Optical Private Network (OPN) backbone [21], see Figure 8, which is composed of dedicated connections between CERN Tier-0 and each of the Tier1s, with the capacity of 10 Gbit/s each. The original connections proliferated into duplicates or backroutes making the system considerably reliable. The OPN is then interconnected with national network infrastructures like the GEANT [22] in Europe or the US-LHCNet [23] and all the National Research and Education Networks (NRENs) in other countries.

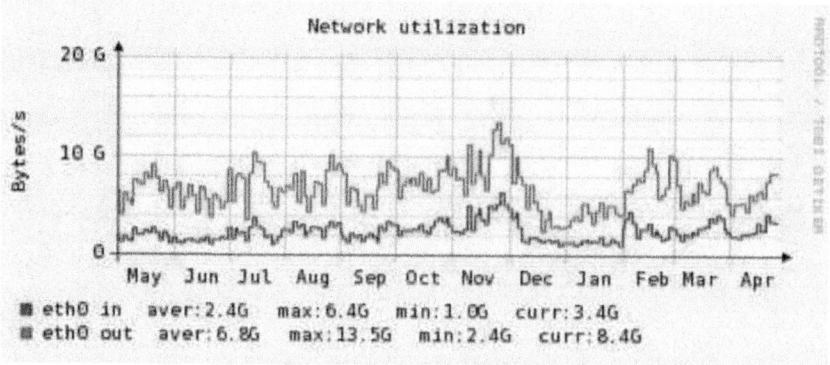

Figure 4: CERN Tier-0 Disk Servers (GB/s), 2010/2011.

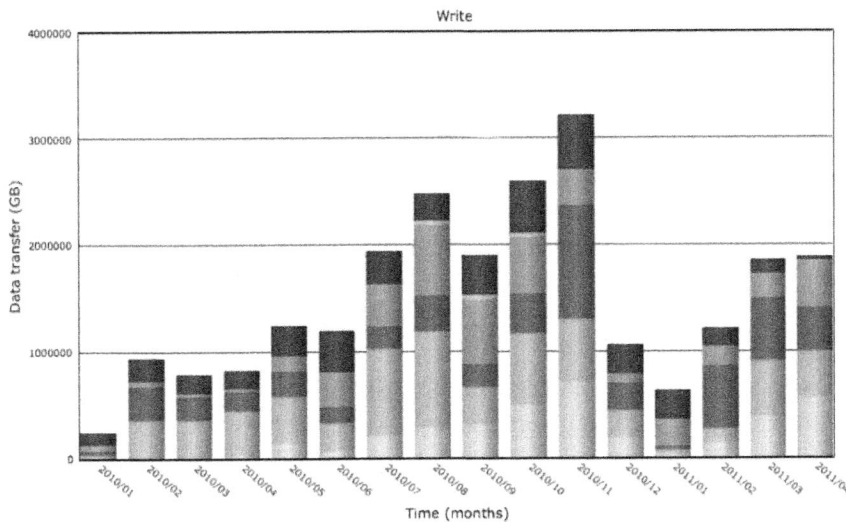

Figure 5: Data written to tape at the CERN Tier-0 (GB/month).

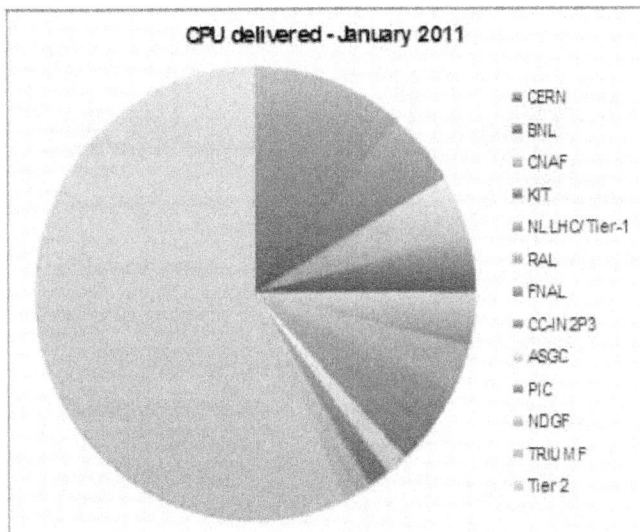

Figure 6: CPU resources in WLCG, January 2011. More than 50% was delivered by Tier-2s.

There exists a concept of so-called LHCONE [24], which should enable a good connectivity of Tier-2s and Tier-3s to the Tier-1s without overloading the general purpose network links. It will extend and complete the existing OPN infrastructure to increase the interoperability of all the WLCG sites.

Data and Service Challenges

As we will describe in section 6, the WLCG data management worked flawlessly when the real data started to flow from the detectors in the end of 2009. This was not just a happy coincidence. There were over 6 years of continuous testing of the infrastructure performance. There was a number of independent experiments' so-called Data Challenges which started in 2004, when the "artificial raw" data was generated in the Monte Carlo productions and then processed and managed as if it was the real raw data. Moreover, there was a series of WLCG Service Challenges also starting in 2004, with the aim to demonstrate WLCG services aspects: data management, scaling of job workloads, security incidents, interoperability, support processes and all was topped with data transfers exercise lasting for weeks. The last test was the Service Challenge STEP'09 including all experiments and testing full computing models. Also, the cosmic data taking which started in 2008 has checked the performance of the data processing chain on a smaller scale.

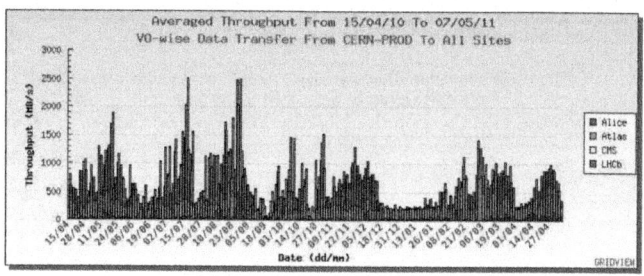

Figure 7: CPU resources in WLCG, January 2011. More than 50% was delivered by Tier-2s.

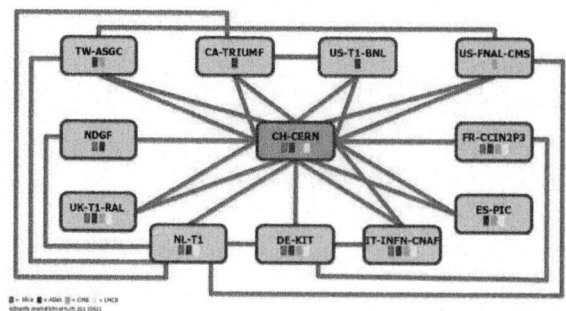

Figure 8: LHCOPN.

Currently, whether the data taking is going on or not, the network, especially the OPN, and the sites are under continuous checking: there are automatically generated test jobs periodically sent over the infrastructure to test the availability and functioning of the network and on-site services.

Concluding Remarks

The WLCG integrates and operates resources distributed all over the world and its task is to make all these resources accessible and usable for the LHC experiments to distribute, archive and process the data produced by the LHC. This task is done using a specialized software called "middleware" because it sits between the operating systems of the computers at the WLCG sites and the Physics applications software used for the reconstruction, analysis and simulation of the LHC data (or any other application software layer). The middleware is a collection of protocols, agents and programs/services which we describe in the following section.

MIDDLEWARE

As we already mentioned, the Worldwide LHC Computing Grid is a distributed computing infrastructure that spans over five continents managing resources distributed across the world (due to funding, operability and access reasons). The resources operated by the WLCG belong either to the two main global grids, EGI [16] and OSG [17], or to other collaborating regional or national grids. To make this diverse variety of resources globally available for all the WLCG users, the WLCG has been developing its own middleware, a software layer that "brings all the resources together": a collection of programs, services and protocols to manage and operate the entire WLCG infrastructure (see Figure 9).

Overview of Grid Services

The WLCG middleware is a complex suite of packages which includes (see also Figure 10):
- Data Management Services:
  - Storage Element
  - File Catalogue Service
  - Grid file access tools
  - File Transfer Service
  - GridFTP service
  - Database and DB Replication Services

    – POOL Object Persistency Service

- Security Services:
  - Certificate Management Service

  - Virtual Organization [25] Management Registration Service (VOMRS)

  - Authentication and Authorization Service (the X509 infrastructure)

- Job Management Services:
  - Compute Element

  - Workload Management

  - Service VO Agent Service

  - Application Software Install Service

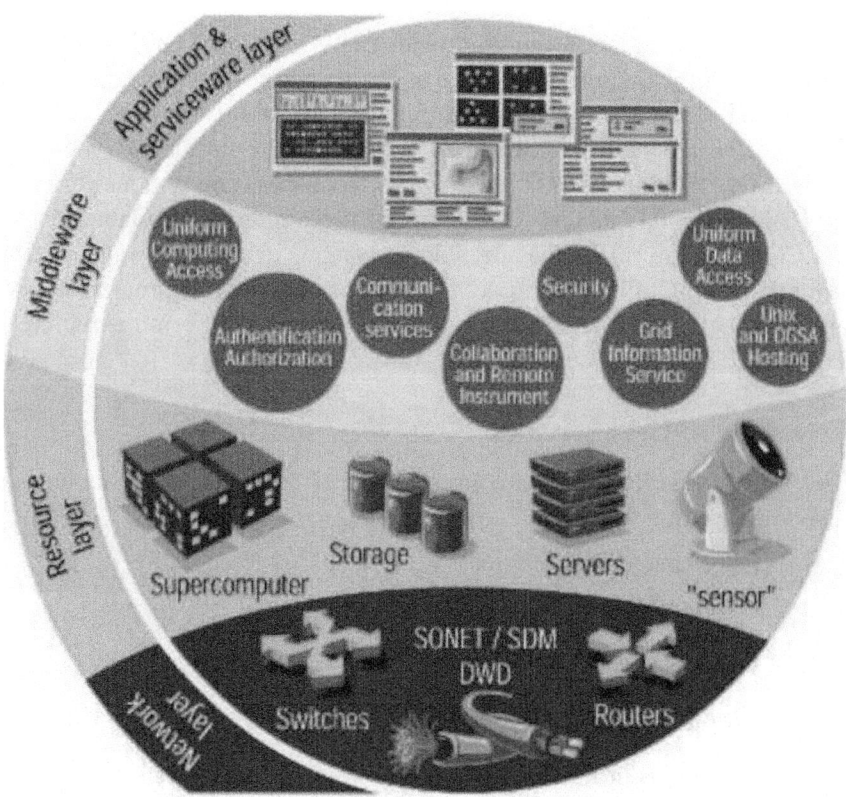

Figure 9: Grid Layers.

- Information Services:
  - Accounting Service
  - Site Availability Monitor
  - Monitoring tools: experiment dashboards; site monitoring

The WLCG middleware has been built and further developed using and developing some packages produced by other projects including, e.g.:

- EMI (European Middleware Initiative) [26], combining the key middleware providers of ARC, gLite, UNICORE and dCache

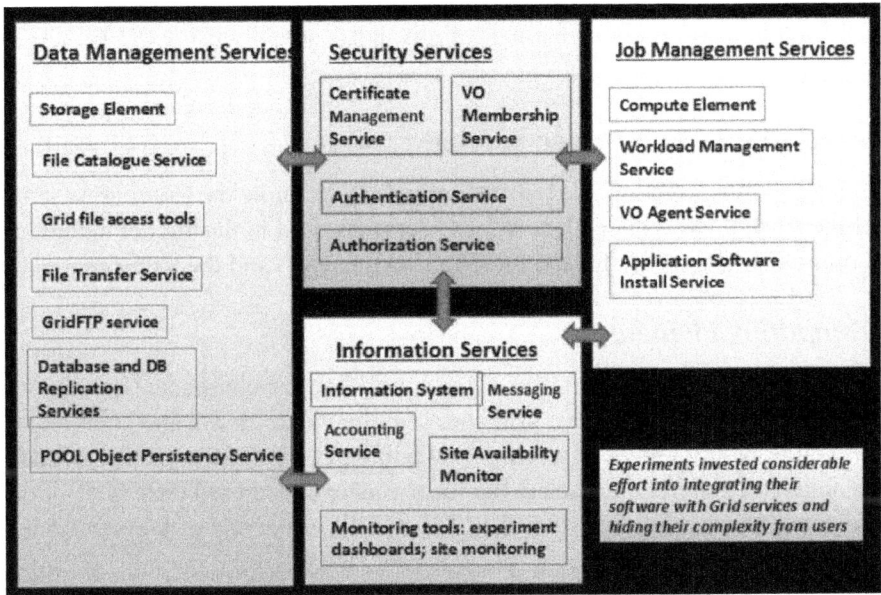

Figure 10: Schema of Grid services.

- Globus Toolkit [27] developed by the Globus Alliance
- OMII from the Open Middleware Infrastructure Institute [28]
- Virtual Data Toolkit [29]

Experiments' Specific Developments

All the LHC experiments created their own specific Computing models summarized in the individual Computing Technical Design Reports (TDRs). They do not rely only on the WLCG-provided middleware packages but are also developing some specific components tailored to better comply with their Computing models.

For example, the ALICE experiment has developed a grid middleware suite AliEn (AliCE Environment [30]), which provides a single interface for a transparent access to computing resources for the ALICE community. AliEn consists of a collection of components and services which will be described in the next section. AliEn, together with selected packages of the WLCG-provided middleware, gives a complete framework to the ALICE community to manage and process the data produced by the LHC according to the ALICE Computing model.

All the LHC experiments invested a considerable effort into shielding the users from the underlying complexity of the Grid machinery, trying to provide relatively simple entry points into the Grid. This effort has payed off and is reflected in a considerable number of physicists actually using the WLCG for their analysis.

Selected WLCG-Provided Services

In the following section, we will describe as an example the Computing model of the ALICE experiment. The WLCG services used in this model include the Computing Element (CE), the Storage Element (SE) and the VOBOX.

Computing Element

The Computing Element (CE) [31] is a middleware component/grid service providing an entry point to a grid site. It authenticates users and submits jobs to Worker Nodes (WN), aggregates and publishes information from the nodes. It includes a generic interface to the local cluster called Grid Gate (GG), Local Resource Management System (LRMS) and the collection of Worker Nodes.

Originally, the submission of jobs to CEs was performed by the Workload Management System (WMS) [32], a middleware component/grid service, that also monitors jobs status and retrieves their output. WLCG (gLite) CE is a computing resource access service using standard grid protocols. To improve the performance, the CREAM (Computing Resource Execution And Management) Computing Element [33] has replaced the gLite-CE in production since about 2009. It is a simple, lightweight service for job management operation at the Computing Element level. CREAM-CE accepts job submission requests (described with the same files as used for the Workload Management System) and other job management requests like, e.g., job monitoring. CREAM-CE can be used by a generic client, e.g., an end-user willing to directly submit jobs to a CREAM-CE, without the WMS component.

Storage Element (XRootD)

The Storage Element (SE) [34] provides storage place and access for data. Important variables apart from available storage space, read/write speeds and bandwidth concern reliability against overload, percentage of failed transfers from/to SE and percentage of lost/corrupted files.

WLCG (gLite) provides dCache [35] and DPM [36] storage management tools used by the LHC experiments. However within the ALICE infrastructure, the preferred storage manager is the Scalla/XRootD package [37] developed within a SLAC [38] - CERN collaboration (originally, it was a common project of SLAC and INFN [39]). After CERN got involved, the XRootD was bundled in ROOT [40] as a generic platform for distributed data access, very well suited for the LHC data analysis.

The primary goal has been the creation of data repositories with no reasonable size limit, with high data access performance and linear scaling capabilities. The framework is a fully generic suite for fast, low latency and scalable data access, which can serve any kind of data, organized as a hierarchical filesystem-like namespace, based on the concept of directory.

"xrootd" is just the name of the data access daemon. Although fundamental, it is just a part of the whole suite. The complete suite is called Scalla/XRootD, Scalla meaning Structured Cluster Architecture for Low Latency Access.

The manager exhibits important features including:

- High speed access to experimental data
- High transaction rate with rapid request dispersement (fast open, low latency)
- Write once read many times processing mode
- Fault tolerance (if servers go, the clients do not die)
- Fault tolerance (able to manage in realtime distributed replicas)
- Integrated in ROOT

From the site administrator point, the following features are important:

- No database requirements (no backup/recovery issues, high performance)
- Resources gentle, high efficiency data server (low CPU/byte overhead, small memory footprint)
- Simple installation
- Configuration requirements scale linearly with site complexity
- No 3rd party software needed (avoids messy dependencies)
- Low administration costs

- Self-organizing servers remove need for configuration changes in big clusters

Additional features:

- Generic Mass Storage System Interface (HPSS, CASTOR, etc)
- Full POSIX access
- Server clustering for scalability, supports large number of clients from a small number of servers
- Up to 262000 servers per cluster
- High WAN data access efficiency (exploit the throughput of modern WANs for direct data access, and for copying files as well)

VOMS

VOMS [41] stands for Virtual Organization Management Service and is one of the most commonly used Grid technologies needed to provide user access to Grid resources. It works with users that have valid Grid certificates and represents a set of tools to assist authorization of users based on their affiliation. It serves as a central repository for user authorization information, providing support for sorting users into a general group hierarchy - users are grouped as members of Virtual Organizations (VOs). It also keeps track of users' roles and provides interfaces for administrators to manage the users. It was originally developed for the EU DataGrid project.

VOBOX

The VOBOX [42] is a standard WLCG service developed in 2006 in order to provide the LHC experiments with a place where they can run their own specific agents and services. In addition, it provides the file system access to the experiment software area. This area is shared between VOBOX and the Worker Nodes at the given site. In the case of ALICE, the VOBOX is installed at the WLCG sites on dedicated machines and its installation is mandatory for sitesto enter the grid production (it is an "entry door" for a site to the WLCG environment). The access to the VOBOX is restricted to the Software Group Manager (SGM) of the given Virtual Organization. Since 2008, this WLCG service has been VOMS-aware [42]. In the following section, we will describe the services running on the VOBOX machines reserved at a site for the ALICE computing.

ALICE COMPUTING MODEL

In this section, we will briefly describe the computing model of the ALICE experiment [43]. ALICE (A Large Ion Collider Experiment) [5] is a dedicated heavy-ion (HI) experiment at the CERN LHC which apart from the HI mission has also its proton-proton (pp) Physics program. Together with the other LHC experiments, ALICE has been successfully taking and processing pp and HI data since the LHC startup in November 2009. During the pp running, the data taking rate has been up to 500 MB/s while during the HI running the data was taken with the rate up to 2.5 GB/s. As was already mentioned, during 2010 the total volume of data taken by all the LHC experiments reached 15 PB, which corresponds to 7 months of the pp running and 1 month of the HI running (together with 4 months of an LHC technical stop for maintenance and upgrades this makes up for one standard data taking year (SDTY)).

The computing model of ALICE relies on the ALICE Computing Grid, the distributed computing infrastructure based on the hierarchical Tier structure as described in section 2. ALICE has developed over the last 10 years a distributed computing environment and its implementation: the Grid middleware suite AliEn (AliCE Environment) [30], which is integrated in the WLCG environment. It provides a transparent access to computing resources for the ALICE community and will be described in the next section.

Raw Data Taking, Transfer and Registration

The ALICE detector consists of 18 subdetectors that interact with 5 online systems [5]. During data taking, the data is read out by the Data Acquisition (DAQ) system as raw data streams produced by the subdetectors, and is moved and stored over several media. On this way, the raw data is formatted, the events (data sets containing information about individual pp or Pb-Pb collisions) are built, the data is objectified in the ROOT [40] format and then recorded on a local disk. During the intervals of continuous data taking called runs, different types of data sets can be collected of which the so-called PHYSICS runs are those substantial for Physics analysis. There are also all kinds of calibration and other subdetectors' testing runs important for the reliable subsystems operation.

ALICE experimental area (called Point2 (P2)) serves as an intermediate storage: the final destination of the collected raw data is the CERN Advanced STORage system (CASTOR) [19], the permanent data storage (PDS) at the CERN Computing center. From Point2, the raw data is transferred to the disk buffer adjacent to CASTOR at CERN (see Figure 11). As mentioned before, the transfer rates are up to 500 MB/s for the pp and up to 2.5 GB/s for the HI data taking periods.

After the migration to the CERN Tier-0, the raw data is registered in the AliEn catalogue [30] and the data from PHYSICS runs is automatically queued for the Pass1 of reconstruction, the first part of the data processing chain, which is performed at the CERN Tier-0. In parallel with the reconstruction, the data from PHYSICS runs is also automatically queued for thereplication to external Tier-1s (see Figure 11). It may happen that the replication is launched and finished fast and the data goes through the first processing at a Tier-1.

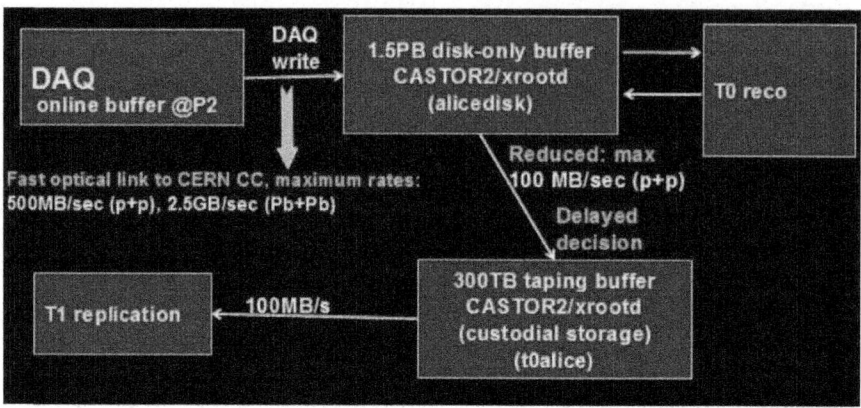

Figure 11: Data processing chain. Data rates and buffer sizes are being gradually increased.

The mentioned automated processes are a part of a complex set of services deployed over the ALICE Computing Grid infrastructure. All the involved services are continuously controlled by automatic procedures, reducing to a minimum the human interaction. The Grid monitoring environment adopted and developed by ALICE, the Java-based MonALISA (MONitoring Agents using Large Integrated Services Architecture) [44], uses decision-taking automated agents for management and control of the Grid services. For monitoring of raw data reconstruction passes see [45].

The automatic reconstruction is typically completed within a couple of hours after the end of the run. The output files from the reconstruction are registered in AliEn and are available on the Grid (stored and accessible within the ALICE distributed storage pool) for further processing.

AliRoot

AliRoot [46] is the ALICE software framework for reconstruction, simulation and analysis of the data. It has been under a steady development since 1998. Typical use cases include detector description, events generation, particle transport, generation of "summable digits", event merging, reconstruction, particle identification and all kinds of analysis tasks. AliRoot uses the ROOT [40] system as a foundation on which the framework is built. The Geant3 [47] or FLUKA [48] packages perform the transport of particles through the detector and simulate the energy deposition from which the detector response can be simulated. Except for large existing libraries, such as Pythia6 [49] and HIJING [50], and some remaining legacy code, this framework is based on the Object Oriented programming paradigm and is written in C++.

AliRoot is constituted by a large amount of files, sources, binaries, data and related documentation. Clear and efficient management guidelines are vital if this corpus of software should serve its purpose along the lifetime of the ALICE experiment. The corresponding policies are described in [51]. For understanding and improvement of the AliRoot performance, as well as for understanding the behavior of the ALICE detectors, the fast feedback given by the offline reconstruction is essential.

Multiple Reconstruction

In general, the ALICE computing model for the pp data taking is similar to that of the other LHC experiments. Data is automatically recorded and then reconstructed quasi online at the CERN Tier-0 facility. In parallel, data is exported to the different external Tier-1s, to provide two copies of the raw data, one stored at the CERN CASTOR and another copy shared by all the external Tier-1s.

For HI (Pb-Pb) data taking this model is not viable, as data is recorded at up to 2.5 GB/s. Such a massive data stream would require a prohibitive amount of resources for quasi real-time processing. The computing model therefore requires that the HI data reconstruction at the CERN Tier-0 and its replication to the Tier-1s be delayed and scheduled for the period of four months of the LHC technical stop and only a small part of the raw data (10-15%) be reconstructed for the quality checking. In reality, comparatively large part of the HI data (about 80%) got reconstructed and replicated in 2010 before the end of the data taking due to occasional lapses in the LHC operations and much higher quality of the network infrastructure than originally envisaged.

After the first pass of the reconstruction, the data is usually reconstructed subsequently more times (up to 6-7 times) for better results at Tier-1s or Tier-2s . Each pass of the reconstruction triggers a cascade of additional tasks organized centrally like Quality Assurance (QA) processing trains and a series of different kinds of analysis trains described later. Also, each reconstruction pass triggers a series of the Monte Carlo simulation productions. All this complex of tasks for a given reconstruction pass is launched automatically as mentioned before.

Analysis

The next step in the data processing chain is then the analysis. There are two types of analysis: a scheduled analysis organized centrally and then the end-user, so-called chaotic analysis. Since processing of the end-user analysis jobs often brings some problems like a high memory consumption (see Figure 12) or unstable code, the scheduled analysis is organized in the form of so-called analysis trains (see [52]). The trains absorb up to 30 different analysis tasks running in succession with one data set read and with a very well controlled environment. This helps to consolidate the end-user analysis.

The computing model assumes that the scheduled analysis will be performed at Tier-1 sites, while the chaotic analysis and simulation jobs will be performed at Tier-2s. The experience gained during the numerous Data Challenges, the excellent network performance, the stable and mature Grid middleware deployed over all sites and the conditions at the time of the real data taking in 2010/2011 progressively replaced the original hierarchical scenario by a more "symmetric" model often referred to as the "cloud model".

Simulations

As already mentioned, ever since the start of building the ALICE distributed computing infrastructure, the system was tested and validated with increasingly massive productionsof Monte Carlo (MC) simulated events of the LHC collisions in the ALICE detector.

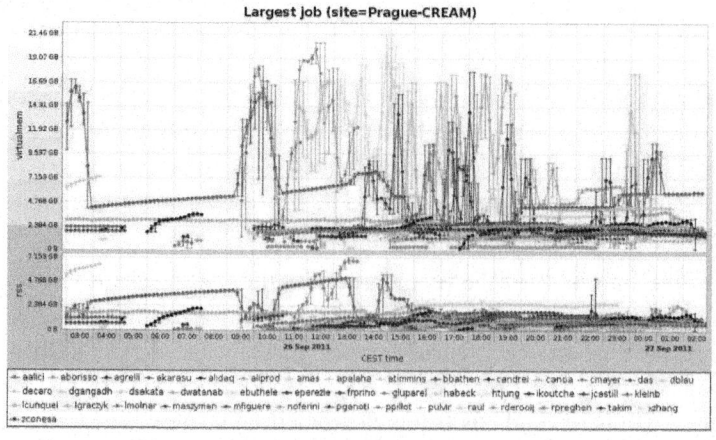

Figure 12: End-user analysis memory consumption: peaks in excess of 20 GB.

The simulation framework [53] covers the simulation of primary collisions and generation of the emerging particles, the transport of particles through the detector, the simulation of energy depositions (hits) in the detector components, their response in form of so-called summable digits, the generation of digits from summable digits with the optional merging of underlying events and the creation of raw data. Each raw data production cycle triggers a series of corresponding MC productions (see [54]). As a result, the volume of data produced during the MC cycles is usually in excess of the volume of the corresponding raw data.

Data Types

To complete the description of the ALICE data processing chain, we will mention the different types of data files produced at different stages of the chain (see Figure 13).

As was already mentioned, the data is delivered by the Data Acquisition system in a form of raw data in the ROOT format. The reconstruction produces the so-called Event Summary Data (ESD), the primary container after the reconstruction. The ESDs contain information like run and event numbers, trigger class, primary vertex, arrays of tracks/vertices, detector conditions. In an ideal situation following the computing model, the EODs should be of 10% size of the corresponding raw data files.

The subsequent data processing provides so-called Analysis Object Data (AOD), the secondary processing product, which are data objects containing more skimmed information needed for final analysis. According to the Computing model, the size of AODs should be 2% of the raw data file size. Since it is difficult to squeeze all the information needed for the Physics results in such small data containers, this limit was not yet fully achieved.

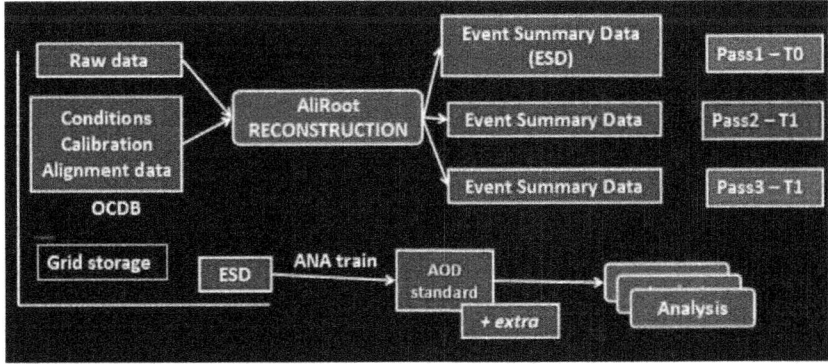

Figure 13: Data types produced in the processing chain.

Resources

The ALICE distributed computing infrastructure has evolved from a set of about 20 computing sites into a global world-wide system of distributed resources for data storage and processing. As of today, this project is made of over 80 sites spanning 5 continents (Africa, Asia, Europe, North and South America), involving 6 Tier-1 centers and more than 70 Tier-2 centers [55], see also Figure 14. Altogether, the resources provided by the ALICEsites represent in excess of 20 thousands of CPUs, 12 PB of distributed disk storage and 30 PB of distributed tape storage, and the gradual upscale of this capacity is ongoing. Similar to other LHC experiments, about half of the CPU and disk resources is provided by the Tier-2 centers. For the year 2012, ALICE plans/requirements for computing resources within WLCG represent 211.7 of kHEP-SPEC06 CPU capacity, 38.8 PB of disk storage and 36.6 PB of tapes [56].

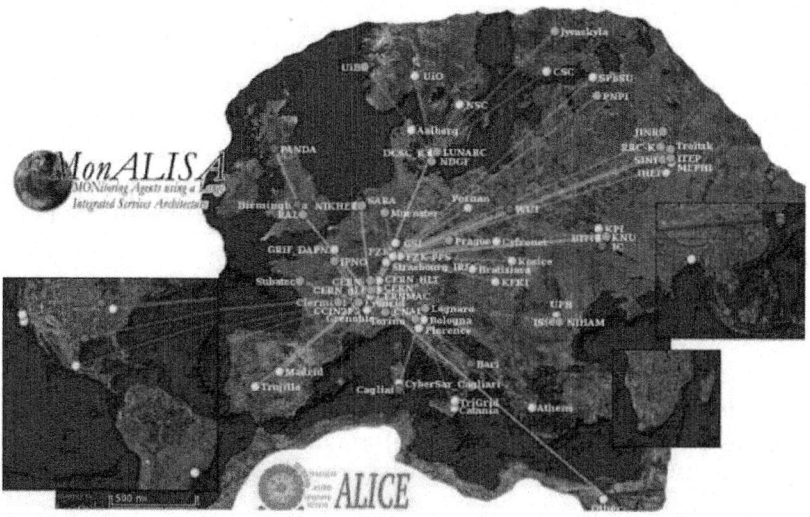

Figure 14: ALICE sites.

Concluding Remarks

The concept of the ALICE computing model was officially proposed in 2005. Since then, it has been used for massive Monte Carlo event productions, for end-user analysis and for the raw data management and processing. The strategy has been validated under heavy load during a series of Data Challenges and during the real data taking in 2010/2011. The model provides the required Grid functionality via a combination of the common Grid services offered on the WLCG resources and the ALICE-specific services from AliEn.

Today's computing environments are anything but static. Fast development in Information Technologies, commodity hardware (hardware being constantly replaced and operating systems upgraded), Grid software and networking technologies inevitably boosted also further development of the ALICE computing model. One of the main effects is a transformation of the model from the strictly hierarchical Tier-like structure to a more loose scenario, a "cloud-like" solution.

AliEn

AliEn [30] is a set of middleware tools and services which represents an implementation of the ALICE distributed computing environment integrated in the WLCG environment. AliEn has been under a constant development by ALICE since 2001 and was deployed over the ALICE Grid infrastructure right from the start. One of the important features is the set of interfaces to other Grid implementations like gLite [57], ARC [58] and OSG [17]. AliEn was initially developed as a distributed production environment for the simulation, reconstruction, and analysis of Physics data. Since it was put in the production in 2001, ALICE has been using AliEn before the start of the real data taking for distributed production cycles of Monte-Carlo simulated raw data, including subsequent reconstruction and analysis, during the regular Physics Data Challenges. Since 2005, AliEn has been used also for end-user analysis. Since December 2007, when the ALICE detector started operation taking cosmic data, AliEn has been used also for management of the raw data. Since the LHC startup in 2009, millions of jobs have been successfully processed using the AliEn services and tools. AliEn developers provided the users with a client/interface - "alien shell" [59] and a set of plugins designed for the end users' job submission and handling. These tools together with the tools provided by the ALICE Grid monitoring framework MonALISA [44], hide the complexity and heterogenity of the underlying Grid services from the end-user while facing the rapid development of the Grid technologies.

- AliEn File Catalogue with metadata capabilities
- Data management tools for data transfers and storage
- Authentication, authorization and auditing services
- Job execution model
- Storage and computing elements
- Information services
- Site services
- Command line interface - the AliEn shell aliensh

- ROOT interface
- Grid and job monitoring
- Interfaces to other Grids

AliEn was primarily developed by ALICE, however it was adopted also by a couple of other Virtual Organizations like PANDA [60] and CBM [61].

File Catalogue (FC)

The File Catalogue is one of the key components of the AliEn suite. It provides a hierarchical structure (like a UNIX File system) and is designed to allow each directory node in the hierarchy to be supported by different database engines, running on different hosts. This building on top of several databases allows to add another database to expand the catalogue namespace and assures scalability of the system and allow growth of the catalogue as the files accumulate over the years.

Unlike real file systems, the FC does not own the files; it is a metadata catalogue on the Logical File Names (LFN) and only keeps an association/mapping between the LFNs and (possibly multiple) Physical File Names (PFN) of real files on a storage system. PFNs describe the physical location of the files and include the access protocol (rfio, xrootd), the name of the AliEn Storage Element and the path to the local file. The system supports file replication and caching. The FC provides also a mapping between the LFNs and Globally Unique Identifiers (GUID). The labeling of each file with the GUID allows for the asynchronous caching.

The write-once strategy combined with GUID labeling guarantees the identity of files with the same GUID label in different caches. It is possible to automatically construct PFNs : to store only the GUID and Storage Index and the Storage Element builds the PFN from the GUID. There are two independent catalogues: LFN->GUID and GUID->PFN. A schema of the AliEn FC is shown in Figure 15.

The FC can also associate metadata to the LFNs. This metadata is a collection of user-defined key value pairs. For instance, in the case of ALICE, the current metadata is the software version used to generate the files, number of events inside a file, or calibration files used during the reconstruction.

Job Execution Model

AliEn's Job execution model is based on the pull architecture. There is a set of central components (Task Queue, Job Optimizer, Job Broker) and another set of site components (Computing Element (CE), Cluster Monitor, MonALISA,

Package Manager). The pullarchitecture has one major advantage with respect to the push one: the system does not have to know the actual status of all resources, which is crucial for large flexible Grids. In a push architecture, the distribution of jobs requires to keep and analyze a huge amount of status data just to assign a job, which becomes difficult in the expanding grid environment.

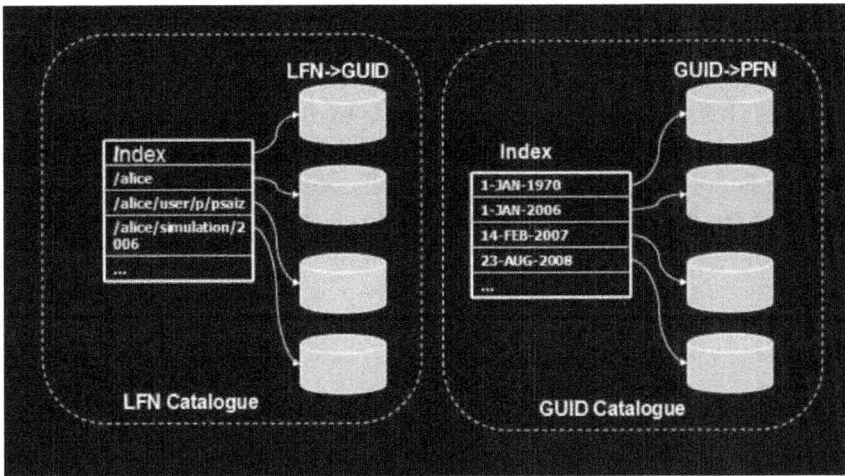

Figure 15: AliEn File Catalogue.

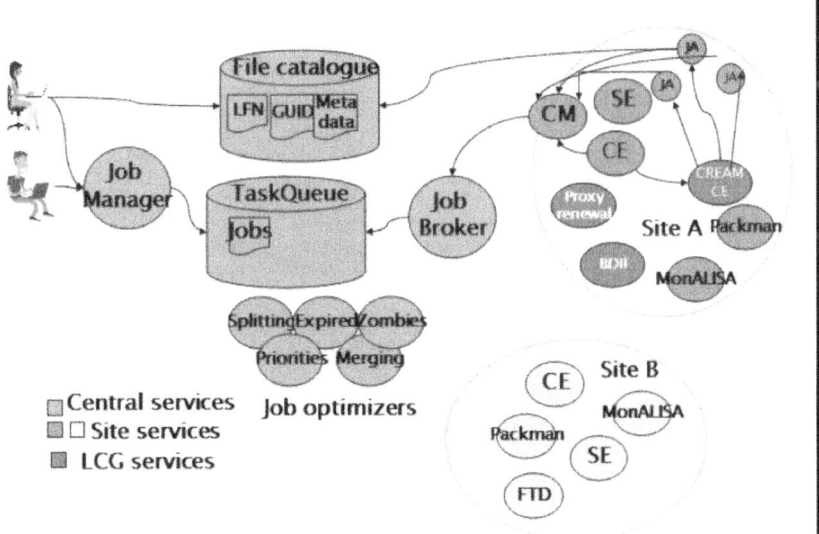

Figure 16: AliEn + WLCG services.

In the pull architecture, local agents (pilot status-checking test jobs) running at individual sites ask for real jobs after having checked the local conditions and found them appropriate for the processing of the job. Thus, AliEn only deals with the requests of local pilot jobs, so-called Job Agents (JA), to assign appropriate real jobs. The descriptions of jobs in the form of ClassAds are managed by the central Task Queue.

Each site runs several AliEn services: CE, ClusterMonitor, Package Manager (PackMan) and a MonALISA client. The AliEn CE automatically generates Job Agents and submits them to the local batch system. The ClusterMonitor manages the connections between the site and central services, so there is only one connection from each site. The AliEn CE can also be submit to the CREAM-CE, ARC, OSG or even the WMS, and delegate the communication with the local batch system to such a service. Schemas of the job submission procedure in AliEn are shown in Figures 16 and 17.

Jobs

When a job is submitted by a user, its description in the form of a ClassAd is kept in the central TQ where it waits for a suitable Job Agent for execution. There are several Job optimizers that can rearrange the priorities of the jobs based on the user quotas. These optimizers can also split jobs, or even suggest data transfers so it would be more likely that some Job Agent picks up the job. After it has been submitted, a job gets through several stages [62].

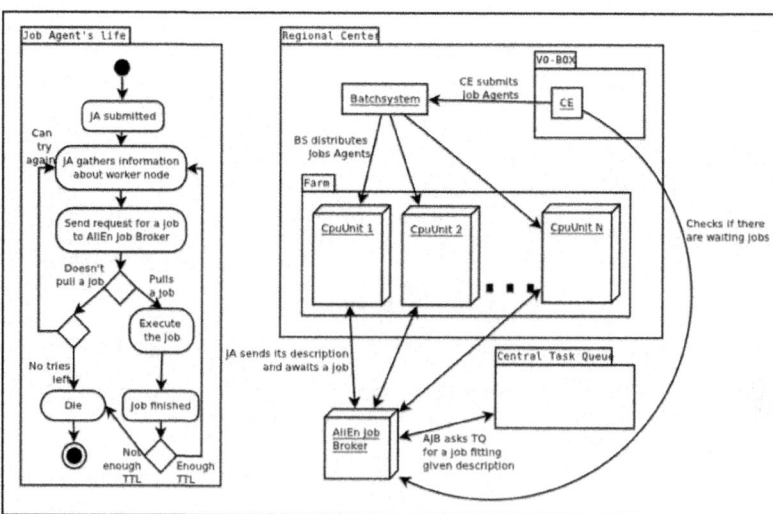

Figure 17: The Job Agent model in AliEn: the JA does five attempts to pull a job before it dies.

The information about running processes is kept also in the AliEn FC. Each job is given a unique id and a corresponding directory where it can register its output. The JAs provide a job-wrapper, a standard environment allowing a virtualization of resources. The whole job submission andprocessing chain is extensively monitored so a user can any time get the information on the status of his/her jobs.

Site Services

As mentioned before, there are several AliEn services running at each ALICE site: CE, ClusterMonitor, PackMan and MonALISA. These services are running on a dedicated machine, so-called VOBOX described in section 3.

The AliEn site CE is usually associated with the local batch system. It is periodically submitting testing pilot jobs (Job Agents) to the local WLCG CE or an appropriate external Resource Broker or WMS. The role of the Job Agents is to verify the local hardware and software capacities at the site. After the usual matchmaking procedure, the JA is sent, through the site CE, into the local batch queue and then to a local Worker Node (WN). After its startup, the JA performs its task and in the case of a positive checkup, the JA requests a "real" job from the central Task Queue via the AliEn Job Broker, or dies otherwise. The PackMan automates the process of installation, upgrades, configuration and removal of the ALICE software packages from the shared software area on the site. It also advertises known/installed packages. The packages are installed on demand, when requested by a Job Agent running on a Worker Node or during the central software deployment over the Grid sites. If a package is not already installed the PackMan would install it along with its dependencies and return a string with commands that client has to execute to configure the package and all its dependencies. The PackMan manages the local disk cache and cleans it, when it needs more space to install newer packages. The Cluster Monitor handles communication with the AliEn Job Broker and gives configuration to JAs. It gets "heartbeats" from the JAs. If it gets no heartbeats from a JA, the existing job will get into the ZOMBIE status (after 1.5 hours) and then it will expire (after 3 hours).

Monitoring

Since the AliEn Workload Management does not depend directly on sophisticated monitoring, no special monitoring tools were developed in AliEn. As the monitoring solution, ALICE has adopted and further developed the Java-based MonALISA framework [44] mentioned already in the previous section. The MonALISA system is designed as an ensemble of autonomous multithreaded, self-describing agent-based subsystems which are registered as dynamic services, and together can collect and process large amounts of information.

The collected monitoring information is published via Web Service for use by AliEn Optimizers or for visualization purposes. An extension of the network simulation code which is a part of MonALISA can provide a tool for optimization and understanding of the performance of the AliEn Grid system.

Storage

Experience with the performance of different types of storage managers shows that the most advanced storage solution is the native XRootD manager [37] described in section 3. It has been demonstrated that with all other parameters being equal (protocol access speed and security) the native XRootD storage clusters exhibit substantially higher stability and availability. The ALICE distributed system of native XRootD clusters is orchestrated by theglobal redirector which allows interacting with the complete storage pool as a unique storage. All storages are on WAN (Wide Area Network).

AliEn Shell - Aliensh

To complete the brief description of AliEn, we mention the client called AliEn shell. It provides a UNIX-shell-like environment with an extensive set of commands which can be used to access AliEn Grid computing resources and the AliEn virtual file system. There are three categories of commands: informative and convenience commands, File Catalogue and Data Management commands and TaskQueue/Job Management commands. The AliEn shell has been created about 4 years ago and become a popular tool among the users for job handling and monitoring.

Concluding Remarks

AliEn is a high-level middleware adopted by the ALICE experiment, which has been used and validated in massive Monte Carlo events production since 2001, in end-user analysis since 2005 and during the real data management and processing since 2007. Its capabilities comply with the requirements of the ALICE computing model. In addition to modules needed to build a fully functional Grid, AliEn provides interfaces to other Grid implementations enabling the true Grid interoperability. The AliEn development will be ongoing in the coming years following the architectural path chosen at the start and more modules and functionalities are envisaged to be delivered.

The Grid (AliEn/gLite/other) services are many and quite complex. Nonetheless, they are working together, allowing to manage thousands of CPUs and PBs of various storage types. The ALICE choice of single Grid Catalogue, single Task Queue with internal prioritization and a single storage access protocol (xrootd) has been beneficial from user and Grid management viewpoint.

WLCG AND ALICE PERFORMANCE DURING THE 2010/2011 LHC DATA TAKING

In this section, we will discuss the experience and performance of the WLCG in general and the ALICE Grid project in particular during the real LHC data taking both during the proton and the lead ion beam periods.

LHC Performance

The LHC delivered the first pp collisions in the end of 2009, and the stable operations startup was in March 2010. Since then, the machine has been working amazingly well compared to other related facilities. Already in 2009, the machine beaten the world record in the beam energy and other records have followed. In 2010, the delivered integrated luminosity was 18.1 pb^{-1} and already during the first months of operation in 2011, the delivered luminosity was 265 pb^{-1} . This is about a quarter of the complete target luminosity for 2010 and 2011 [63], which is supposed to be sufficient to get the answer concerning the existence of the Higgs boson. Also, as mentioned in section 1, the machine has beaten the records concerning the stored energy and also the beam intensity.

The target number of bunches per a beam of protons stated for 2011 was reached already in the middle of the year: 1380 bunches. The final target luminosity of 1034 cm^{-2} s $^{-1}$ is being approached rapidly, at present it is about 1033 cm^{-2} s $^{-1}$ [64].

WLCG Performance

The performance of the data handling by the WLCG has also been surprisingly good. This wrapped up several years to the LHC startup, when the WLCG and the experiments themselves were regularly performing a number of stress-tests of their distributed computing infrastructure and were gradually upgrading the systems using new technologies. As a result, when the data started to flow from the detectors, the performance of the distributed data handling machinery was quite astounding. All aspects of the Grid have really worked for the LHC and have enabled the delivery of Physics results incredibly quickly.

During 2010, 15 PetaBytes of data were written to tapes at the CERN Tier-0 reaching the level expected from the original estimates for the fully-tuned LHC operations, a nominal data taking year. As mentioned in section 2, in average 2 PB of data per a month were written to tapes at Tier-0 with the exception of the heavy ions period when this number got about doubled and a world record was reached with 225 TB written to tapes within one day, see Figure 18. The CERN Tier-0 moved altogether above 1 PB of data per day.

As mentioned in section 2, the mass storage system at the CERN Tier-0 supported data rates at an average over the year of 2.5 GB/s IN with peaks up to 11 GB/s, and data was served at an average rate of $\sim$ 7 GB/s with peaks up to 25 GB/s.

The data processing went on without basic show-stoppers. The workload management system was able to get about 1 million of jobs running per day, see Figure 19, and this load is gradually going up. This translates into significant amounts of computer time. Towards the end of 2010 there was a situation when all of the available job slots at Tier-1s and Tier-2s were often fully occupied. This has been showing up also during 2011, so the WLCG collaboration has already now fully used all the available computing resources. During 2010, the WLCG delivered about 100 CPU-millennia.

As also briefly mentioned in section 2, the WLCG was very successful concerning the number of individuals really using the grid to perform their analysis. At the start of the project, there was a concern that end users will be discouraged from using the grid due to complexity of its structure and services. But thanks to the effort of the WLCG and experiments themselves a reasonably simple access interfaces were developed and the number of end users reached up to 800 in the large experiments.

The distribution of the delivered CPU power between sites has been basically according to the original design, but the Tier-2s provided more than the expected 40%: it was in fact 50% or more of the overall delivery, see section 2. Member countries pledged different amounts of resources according to their capacities and have been delivering accordingly. So the concept of collaborative Grid resource sharing really works and enables institutes worldwide to share data, and provide resources to the common goals of the collaboration.

Network Performance

The key basis for building up a distributed system is the data transfer infrastructure. The network which the WLCG operates today is much in advance of what was anticipated in the time of writing the WLCG TDR.

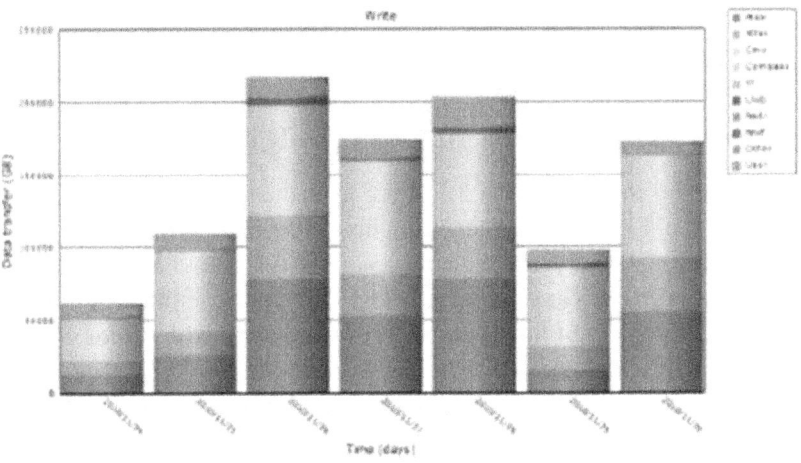

Figure 18: A record in data tape recording: over 220 TB/day.

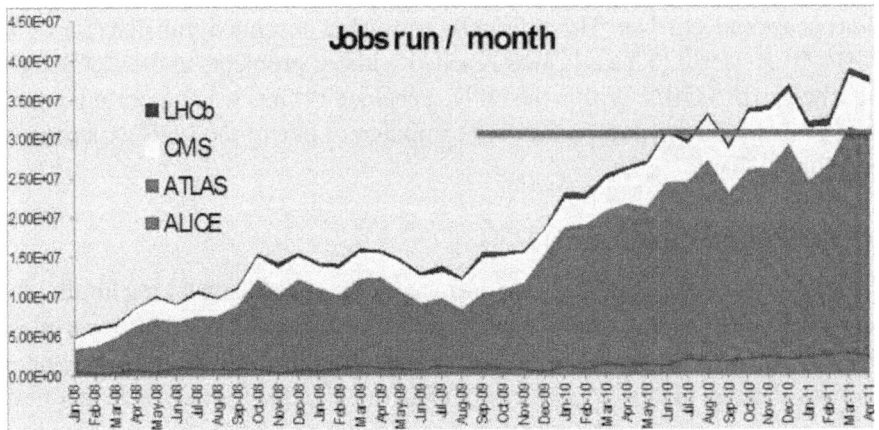

Figure 19: WLCG job profile.

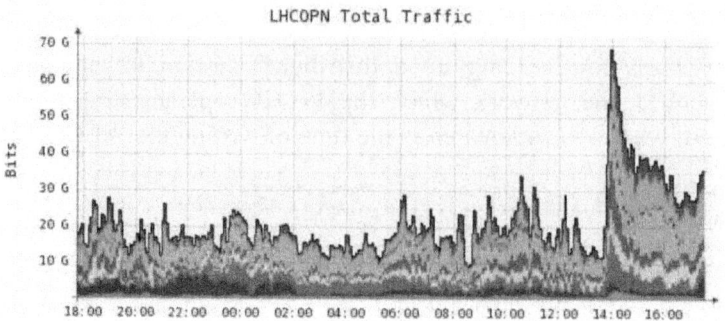

Figure 20: WLCG OPN traffic in 2010 with a peak of 70 Gbit/s.

The OPN, see also section 2, started as dedicated fiber links from CERN to each of the Tier-1s with the throughput 10 Gbit/s. Today, there is a full redundancy in this network with the original links doubled and with back-up links between Tier-1s themselves. The OPN is a complicated system with many different layers of hardware and software and getting it into the current shape was a difficult task, which evidently paid-off.

The original concerns about the possible network unreliability and insufficiency were not realized. The network infrastructure relying on the OPN and the complementary GEANT, US-LHCNet and all the R&E national network infrastructures, extensively monitored and continuously checked with the test transfer jobs, has never been a problem in the data transfer except for occasional glitches. The originally estimated sustained transfer rate of 1.3 GB/s from Tier-0 to Tier-1s was reached without problems and exceeded and reached up to 5 GB/s. Within the OPN, a peak of 70 Gb/s was supported without any problem during a re-processing campaign of one of the LHC experiments, see Figure 20.

Concluding Remarks – WLCG

The experience from the first year and a half of the LHC data taking implies that the WLCG has built a truly working grid infrastructure. The LHC experiments have their own distributed models and have used the WLCG infrastructure to deliver Physics results within weeks after the data recording which has never been achieved before. The fact that a significant numbers of people are doing analysis on the Grid, that all the resources are being used up to the limits and the scientific papers are produced with an unprecedented speed is proving an evident success of the WLCG mission.

ALICE Performance

To conclude this section, we will briefly summarize the experience and performance of the ALICE experiment. ALICE started extremely successfully the processing of the LHC data in 2009: the data collected during the first collisions delivered by the LHC on November 23rd (2009) got processed and analyzed so fast that within one week the article with the results from the first collisions was accepted for publication as the first ever scientific paper with the Physics results from the LHC collisions [65].

Jobs

During the data taking in 2010, ALICE collected 2.3 PB of raw data, which represented about 1.2 million of files with the average file size of 1.9 GB. The data processing chain has been performing without basic problems. The Monte Carlo simulation jobs together with the raw data reconstruction and organized analysis (altogether the organized production) represented almost 7 millions of successfully completed jobs, which translates into 0.3 jobs/second. The chaotic (end user) analysis made for 9 millions of successfully completed jobs, which represents 0.4 jobs/s, consuming approximately 10% of the total ALICE CPU resources (the chaotic analysis jobs are in general shorter than the organized processing jobs). In total, there were almost 16 millions of successfully done jobs, which translates to 1 job/s and 90 thousands jobs/day. The complimentary number of jobs which started running on the Grid but finished with an error was in excess of this.

The running jobs profile got in peaks to 30 thousands of concurrently running jobs (see Figure 21) with more than 50% of the CPU resources delivered by the Tier-2 centers. About 60% of the total number of jobs represented the end user analysis (see Figure 22). In general, the user analysis already in 2010 was a resounding success, with almost 380 people actively using the Grid. Since the chaotic analysis brings sometimes problems concerning not completely perfect code resulting, e.g., in a high memory consumption (cf. section 4), ALICE was running a mixture of the organized production and end user jobs at all its sites, and this scenario was working well.

Storage-2010

The distributed storage system endured and was supporting an enormous load. During 2010/2011, 25.15 PB of data (raw, ESDs, AODs, Monte Carlo productions) was written to xrootd Storage Elements with the speed maximum of 621.1 MB/s. 59.97 PB of data was read from the xrootd Storage Elements, with the speed maximum of 1.285 GB/s, see Figure 23.

Data Taking 2011

Constantly upgrading and extending its hardware resources and updating the grid software, ALICE continued a successful LHC data handling campaign in 2011. By September, the total volume of the collected raw data was almost 1.7 PB with the first reconstruction pass completed. The year 2011 was marked by massive user analysis on the Grid. In May, the most important conference in the Heavy Ion Physics community, Quark Matter 2011 (QM2011) [66], took place and was preceded by an enormous end user analysis campaign. In average, 6 thousands end-user jobs were running all the time, which represents almost 30% of the CPU resources officially dedicated to ALICE (The number of running jobs is higher than that most of the time due to use of opportunistic resources). During the week before the QM2011, there was a peak with 20 thousands of concurrently running end-user jobs, see Figures 23,24. The number of active Grid users reached 411.

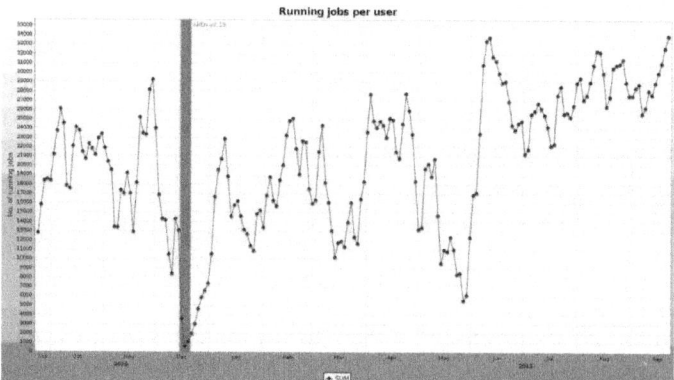

Figure 21: ALICE running jobs profile 2010/2011.

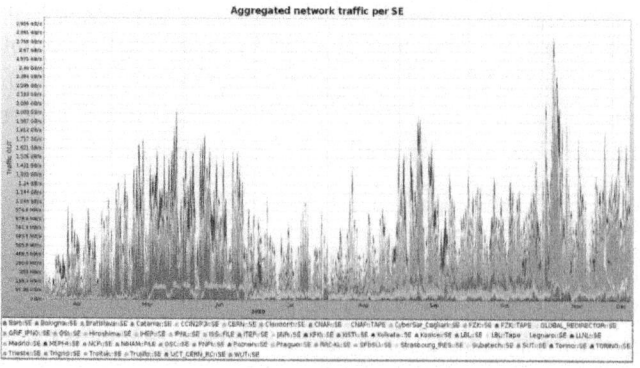

Figure 22: Network traffic OUT by analysis jobs – 2010.

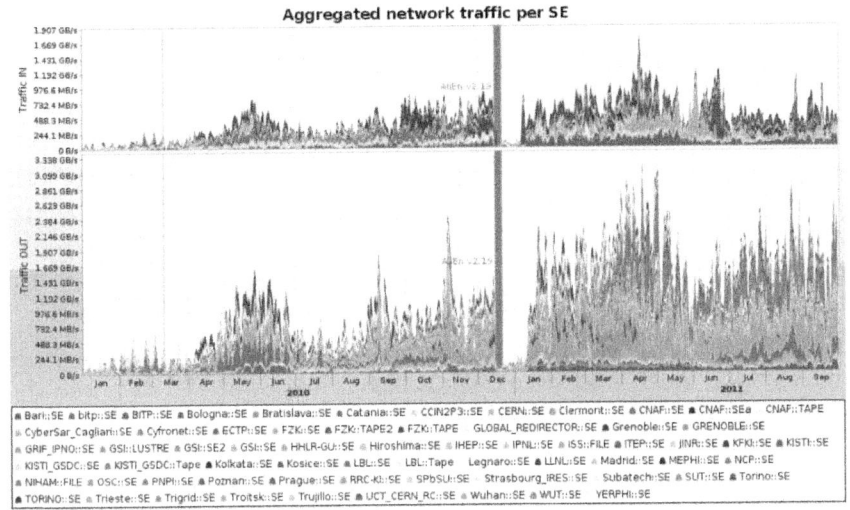

Figure 23: Total network traffic at the ALICE Storage Elements - 2010/2011.

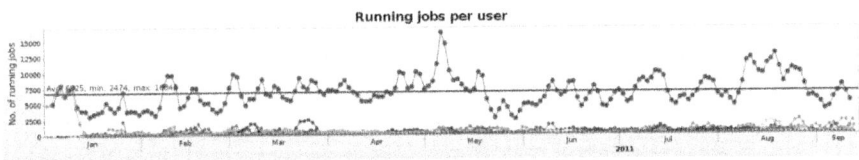

Figure 24: End-user jobs profile – 2011.

In total, the ALICE sites were running in average 21 thousands of jobs with peaks up to 35 thousands (Figure 21). The resources ratio remained 50% delivered by Tier-0 and Tier-1s to 50% delivered by Tier-2s. Altogether, 69 sites were active in the operations. The sites' availability and operability kept very stable throughout the year. The gLite (now EMI) middleware, see section 3, is mature and only a few changes are necessary.

In the beginning of the 2011 campaign, there was a concern that the storage would be saturated. In fact the storage infrastructure was performing basically without problems supporting the enormous load from the end-user analysis and was getting ready for the Pb-Pb operations. The network situation, as was already mentioned for the WLCG in general, has been excellent and allowed for the operation scenario where the hierarchical tiered structure got blurred, the sites of all levels were well interconnected and running a similar mixture of jobs. As a result, the ALICE Grid in a sense has been working as a cloud.

Concluding Remarks – ALICE

In general, similar to the overall characteristics of the WLCG performance also the ALICE operations and data handling campaigns were notably successful right from the beginning of the LHC startup, making the Grid infrastructure operational and supporting a fast delivery of Physics results. By September 2011, ALICE has published 15 scientific papers with the results from the LHC collisions and more is on the way. Two papers [67,68] were marked as "Suggested reading" by the Physical Review Letters editors and the later was also selected for the "Viewpoint in Physics" by Physical Review Letters.

The full list of the ALICE papers published during 2009-2011 can be found on [69]. One of the Pb-Pb collision events recorded by ALICE is shown on Figure 25.

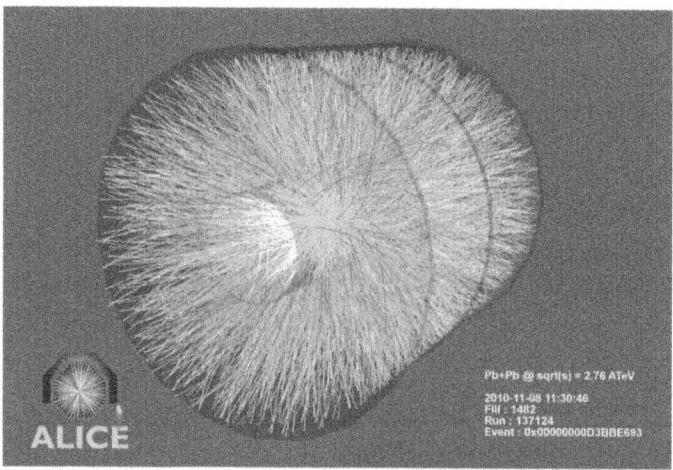

Figure 25: Pb-Pb collision event recorded by ALICE.

SUMMARY AND OUTLOOK

This chapter is meant to be a short overview of the facts concerning the Grid computing for HEP experiments, in particular for the experiments at the CERN LHC. The experience gained during the LHC operations in 2009-2011 has proven that for this community, the existence of a well performing distributed computing is necessary for the achievement and fast delivery of scientific results. The existing WLCG infrastructure turned up to be able to support the data production and processing thus fulfilling its first-plan mission. It has been and will be continuously developing into the future absorbing and giving rise to new technologies, like the advances in networking, storage systems and inter-operability between Grids and Clouds [70,71].

Data Management

Managing the real data taking and processing in 2009-2011 provided basic experience and a starting point for new developments. The excellent performance of the network which was by far not anticipated in the time of writing the WLCG (C)TDR shifted the original concept of computing models based on hierarchical architecture to a more symmetrical mesh-like scenario. In the original design, the jobs are sent to sites holding the required data sets and there are multiple copies of data spread over the system due to anticipation that network will be unreliable or insufficient. It turned out that some data sets were placed on sites and never touched.

Based on the existing excellent network reliability and growing throughput, the data models start to change along a dynamical scenario. This includes sending data to a site just before a job requires it, or reading files remotely over the network, use remote (WAN) I/O to the running processes. Certainly, fetching over the network one needed data file from a given data set which can contain hundreds of files is more effective than a massive data sets deployment and will spare storage resources and bring less network load.

The evolution of the data management strategies is ongoing. It goes towards caching of data rather than strict planned placement. As mentioned, the preferences go to fetching a file over the network when a job needs it and to a kind of intelligent data pre-placement. The remote access to data (either by caching on demand and/or by remote file access) should be implemented.

Network

To improve the performance of the WLCG-operated network infrastructure, the topology of LHC Open Network Environment (LHCONE [24]) is being developed and built. This should be complementary to the existing OPN infrastructure providing the inter-connectivity between Tier-2s and Tier-1s and between Tier-2s themselves without putting an additional load on the existing NREN infrastructures. As we learned during the last years, the network is extremely important and better connected countries do better.

Resources

During the 2010 data taking the available resources were sufficient to cover the needs of experiments, but during 2011 the computing slots as well as the storage capacities at sites started to be full. Since the experience clearly shows that delivery of the Physics results is limited by resources, the experiments are facing a necessity of more efficient usage of existing resources. There are task forces studying the possibility of using the next generations computing and storage technologies. There is for instance a question of using multicore processors which might go into the high performance computing market while WLCG prefers usage of commodity hardware.

Operations

Another important issue is sustainability and support availability for the WLCG operations. The middleware used today for the WLCG operations is considerably complex with many services unnecessarily replicated in many places (like, e.g., databases) mainly due to original worries concerning network. The new conception is to gradually search for more standard solutions instead of often highly specialized middleware packages maintained and developed by WLCG.

Clouds and Virtualization

Among the new technologies, the Clouds is the right buzzword now and the virtualization of resources comes along. The virtualization of WLCG sites started prior to the first LHC collisions and has gone quite far. It helps improving system management, provision of services on demand, can make use of resources more effective and efficient. Virtualization also enables to make use of industrial and commercial solutions.

But, no matter what the current technologies advertise, the LHC community will always use a Grid because the scientists need to collaborate and share resources. No matter what technologies are used underneath the Grid, the collaborative sharing of resources and the network of trust and all the security infrastructure developed on the way of building the WLCG is of enormous value, not only to WLCG community but to e-science in general. It allows people to collaborate across the infrastructures.

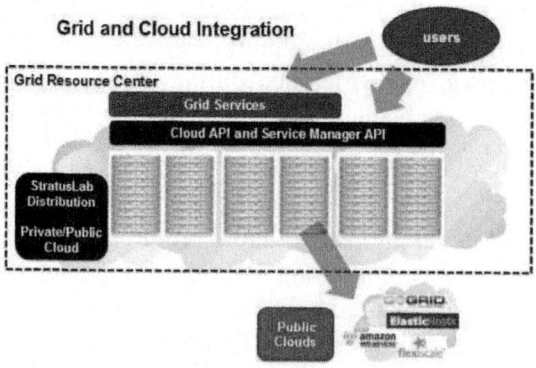

Figure 26: Schema of StratusLab IaaS Cloud interoperability with a Grid.

The basic operations like distributed data management, the high data throughput and the remote job submission can probably be more cloud-like. There is a great interest among people to use commercial Clouds resources, especially when the experiments see their resources becoming full. So, can we use Amazon or Google to do processing of data from LHC? The point is, one cannot be sure what the level of services will be and what the IN/OUT bandwidth will be. This can in principle be negotiated with these companies and may bring some level of agreement. That in principle is doable.

Grids and Clouds

As argued in [70], "Cloud Computing not only overlaps with Grid Computing, it is indeed evolved out of Grid Computing and relies on Grid Computing as its backbone and infrastructure support. The evolution has been a result of a shift in focus from an infrastructure that delivers storage and compute resources (in the case of Grids) to one that is economy based ....". Both the Grids and the Clouds communities are facing the same problems like the need to operate large facilities and to develop methods by which users/consumers discover, request and use resources provided by the centralized facilities.

There exist a number of projects looking into and developing Cloud to Grid interfaces with the idea that Grid and Cloud Computing serve different use cases and can work together improving Distributed Computing Infrastructures (see, e.g., [71]). Also CERN is involved in this activity together with other international laboratories in Europe. With the WLCG resources becoming used up to their limits, using commercial Clouds to process the LHC data is a strategy that should be assessed. Several of the LHC experiments have done tests whether they can use commercial Clouds. But today, the cost is rather high. Also, there are issues like whether academic data can be shipped through academic networks to a commercial provider or how to make sure what happens to this data.

Nevertheless the strategy towards deployment over the WLCG resources Cloud interfaces, managed with high level of virtualization, is under evaluation. Some level of collaboration with industry would provide the understanding how to deploy this properly and what would be the cost. The Cloud and Grid interfaces can be deployed in parallel or on top of each other. This development might also give a way to evolve into a more standardized infrastructure and allow to make a transparent use of commercial Clouds. A testbed of such an architecture is the CERN LXCloud [72] pilot cluster. Implementation at CERN allows to present a Cloud interface or to access other public or commercial Clouds.

This is happening with no change to any of the existing Grid services. Another interesting example is the development of a comprehensive OpenSource IaaS (Infrastructure as a Service) Cloud distribution within the StratusLab project [71], see Figure 26. Anyone can take the code and deploy it on his site and have IaaS Cloud running on his site. The project is focused on deploying Grid services on top of this Cloud, 1) to be a service to existing European Grid infrastructures and to enable these people to use Cloud-operated resources and 2) because the developers consider the Grid services very complex and making sure they run safely on this Cloud should guarantee that also other applications will run without problems.

Physics Results Achieved by the LHC Experiments

Before we bring the final concluding remarks for our chapter, we will briefly summarize the Physics results delivered by the LHC experiments by the time of writing this document. In addition to many specific new results describing different Physics phenomena in the energy regime never explored before, there have been new findings concerning some of the major issues addressed by the LHC research.

- ATLAS and CMS experiments have been delivering results concerning the energy regions excluding the mass of the Higgs boson. The latest results on the topic of Higgs boson searches exclude a wide region of Higgs boson masses: ATLAS excludes Higgs boson masses above 145 GeV, and out to 466 GeV (apart from a couple of points in-between, which are however excluded by CMS studies). For some of the latest references see [73-75].

- To contribute new results on the topic of the dominance of matter over antimatter in the present Universe, the LHCb experiment has been pursuing studies of phenomena demonstrating the so-called CP-symmetry violation. Violation of this symmetry plays an important role in the attempts of Cosmology to explain the dominance of matter over antimatter in our world. The latest LHCb results concerning the demonstration of the existence of the CP-violation can be found, e.g., in [76].

- The study of properties of the Quark Gluon Plasma (QGP), the phase of matter which existed in a fraction of a second after the Big Bang, is the mission of the ALICE experiment. During the lead-lead collisions at the LHC energies, the individual collisions can be seen as "little Big Bangs". The matter produced in these collisions is under extreme conditions: the energy density corresponds to a situation when 15 protons are squeezed into the volume of one proton and the temperature reaches more than 200000 times the temperature in the core of the Sun. ALICE has confirmed the previous findings of the STAR experiment at the Relativistic Heavy Ion Collider (RHIC) at Brookhaven that this QGP behaves like an ideal liquid [68] even at the LHC energies.

Concluding Remarks

As we already stressed, the WLCG performance during the LHC data taking in 2009-2011 was excellent and the basic mission of the WLCG has been fulfilled: the data taking and processing is ongoing without major show-stoppers, hundreds of people are using the Grid to perform their analysis and unique scientific results are delivered within weeks after the data was recorded. In addition, the experience gained during this data taking .stress test. launched new strategies to be followed on the way of the future WLCG development. There are fundamental issues like the approaching lack of WLCG resources and the expansion of new technologies like the Cloud computing. In the time of writing this chapter it looks like we will see in the future some combination of Grid and Cloud technologies will be adopted to operate the distributed computing infrastructures used by the HEP experiments.

ACKNOWLEDGEMENTS

We would like to thank Jiri Adam for a critical reading of the manuscript and for a great help with the LATEX matters. The work was supported by the MSMT CR contracts No. 1P04LA211 and LC 07048.

REFERENCES

1.   D.H. Perkins: Introduction to High Energy Physics, Cambridge University Press, 4th edition (2000), ISBN-13: 978-0521621960. Procedings of 35th International Conference of High Energy Physics, July 22-28, 2010, Paris, France, Proceedings of Science (PoS) electronic Journal: ICHEP 2010

2.   STAR Collaboration: Experimental and theoretical challenges in the search for the quark gluon plasma: The STAR Collaboration's critical assessment of the evidence from RHIC collisions, Nucl. Phys. A757 (2005) 102-183.

3.   CERN - the European Organization for Nuclear Research; http://public. web.cern.ch/public/

4.   The Large Hadron Collider at CERN; http://lhc.web.cern.ch/lhc/; http:// public.web.cern.ch/public/en/LHC/LHC-en.html

5.   ALICE Collaboration: http://aliceinfo.cern.ch/Public/Welcome.html

6. ATLAS Collaboration: http://atlas.ch/

7. CMS Collaboration: http://cms.web.cern.ch/

8. LHCb Collaboration: http://lhcb-public.web.cern.ch/lhcb-public/

9. TOTEM Experiment: http://cern.ch/totem-experiment

10. LHCf Experiment: http://cdsweb.cern.ch/record/887108/files/ lhcc-2005-032.pdf

11. W.N. Cottingham and D.A. Greenwood: An Introduction to the Standard Model of Particle Physics, Cambridge University Press, 2nd edition (2007), ISBN-13: 978-0521852494

12. Worldwide LHC Computing Grid: http://public.web.cern.ch/public/en/lhc/Computing-en.html

13. LHC Computing Grid: Technical Design Report, http://lcg.web.cern.ch/LCG/tdr/

14. I. Foster and C. Kesselman. The Grid: Blueprint for a New Computing Infrastructure. Morgan Kaufmann, 1999; I. Foster et al: The Anatomy of the Grid: Enabling Scalable Virtual Organizations, International Journal of High Performance Computing Applications Vol.15(2001), p.200.

15. WLCG Memorandum of Understanding, http://lcg.web.cern.ch/lcg/mou.htm

16. EGI - The European Grid Initiative; http://web.eu-egi.eu/

17. OSG - The Open Science Grid, http://www.opensciencegrid.org/; https://osg-ress-1.fnal.gov:8443/ReSS/ReSS-prd-History.html

18. I. Legrand et al: MONARC Simulation Framework, ACAT'04, Tsukuba, Japan,2004; http://monarc.cacr.caltech.edu:8081/www_monarc/monarc.htm

19. CERN Advanced Storage Manager: http://castor.web.cern.ch/castor/

20. I. Bird: LHC Computing: After the first year with data, TERENA Networking Conference (TNC2011), Prague, 2011, https://tnc2011.terena.org/web/media/archive/7A

21. LHCOPN - The Large Hadron Collider Optical Private Network, https://twiki.cern.ch/twiki/bin/view/LHCOPN/WebHome

22. The GEANT Project, http://archive.geant.net/

23. USLHCNet: High speed TransAtlantic network for the LHC community, http://lhcnet.caltech.edu/

24. LHCONE-LHC Open Network Environment: http://lhcone.net/

25. Virtual Organisation: http://technical.eu-egee.org/index.php?id=147

26. The European Middleware Initiative: http://www.eu-emi.eu/home

27. The Globus Toolkit: http://www-unix.globus.org/toolkit/

28. OMII - The Open Middleware Infrastructure Institute: http://www.omii. ac.uk/

29. The Virtual Data Toolkit, http://vdt.cs.wisc.edu/

30. P. Saiz et al., AliEn-ALICE environment on the GRID, Nucl. Instrum. Meth. A502 (2003) 437; http://alien2.cern.ch/

31. Computing Element: http://glite.cern.ch/lcg-CE/

32. Workload Management System: http://glite.cern.ch/glite-WMS/

33. The CREAM (Computing Resource Execution And Managemen) Service, http://glite.cern.ch/glite-CREAM/

34. Storage Element: http://glite.cern.ch/glite-SE_dpm_mysql/

35. dCache: http://www.dcache.org/

36. The Disk Pool Manager: https://www.gridpp.ac.uk/wiki/Disk_Pool_Manager; https://twiki.cern.ch/twiki/bin/view/LCG/DataManagementTop

37. XRootD: http://project-arda-dev.web.cern.ch/project-arda-dev/xrootd/site/index.html

38. SLAC (Stanford Linear Accelerator Center): http://slac.stanford.edu/

39. INFN - The National Institute of Nuclear Physics: http://www.infn.it/indexen.php

40. R. Brun and F. Rademakers, ROOT: An object oriented data analysis framework, Nucl. Instrum. Meth. A389 (1997) 81; http://root.cern.ch

41. VOMS-Virtual Organization Membership Service: http://glite.web.cern.ch/glite/packages/R3.1/deployment/ glite-VOMS_mysql/glite-VOMS_mysql.asp

42. The VO-box: http://glite.cern.ch/glite-VOBOX/

43. ALICE Experiment Computing TDR: http://aliceinfo.cern.ch/Collaboration/Documents/TDR/Computing. html

44. Monitoring Agents using a Large Integrated Services Architecture: http://monalisa.cern.ch/monalisa.html; C. Grigoras et al., Automated agents for management and control of the ALICE Computing Grid, Proceedings of the 17th Int. Conf. CHEP 2009, Prague, March 21-27, 2009, J. Phys.: Conf. Ser. 219, 062050.

45. ALICE raw data production cycles: http://alimonitor.cern.ch/production/raw.jsp

46. AliRoot: http://aliceinfo.cern.ch/Offline/AliRoot/Manual.html

47. GEANT3-Detector Description and Simulation Tool: http://wwwasd. web.cern.ch/wwwasd/geant/

48. FLUKA-Particle Physics MonteCarlo Simulation package: http://www. fluka.org/fluka.php

49. Pythia6: http://projects.hepforge.org/pythia6/; Pythia8: http://home.thep. lu.se/~torbjorn/pythiaaux/present.html

50. Xin-Nian Wang and Miklos Gyulassy: HIJING: A Monte Carlo model for multiple jet production in pp, pA and AA collisions, Phys. Rev. D44 (1991), 3501; http://www-nsdth.lbl.gov/~xnwang/hijing/

51. ALICE Offline policy: http://aliceinfo.cern.ch/Offline/General-Information/ Offline-Policy.html

52. Monitoring of Analysis trains in ALICE: http://alimonitor.cern.ch/prod/

53. ALICE simulation framework: http://aliceinfo.cern.ch/Offline/Activities/ Simulation/index. html

54. ALICE MC simulation cycles: http://alimonitor.cern.ch/job_details.jsp

55. ALICE Computing sites: http://pcalimonitor.cern.ch:8889/reports/; ALICE Distributed Storage: http://pcalimonitor.cern.ch/stats?page=SE/ table

56. Yves Schutz: Computing resources 2011-2013, ALICE Computing Board Sept. 1st, 2011: http://indico.cern.ch/materialDisplay.py?contribId= 3&materialId=2&confId=153622; https://twiki.cern.ch/twiki/bin/view/ FIOgroup/TsiBenchHEPSPEC

57. gLite-Lightweight Middleware for Grid Computing, http://glite.cern.ch/

58. ARC-The Advanced Resource Connector middleware, http://www. nordugrid.org/arc/about-arc.html

59. The AliEn Shell-aliensh, AliEn User Interfaces: http://project-arda-dev. web.cern.ch/project-arda-dev/alice/   apiservice/guide/guide-1.0.html#_ Toc156731986; ALICE Grid Analysis: http://project-arda-dev.web.cern. ch/project-arda-dev/alice/ apiservice/AA-UserGuide-0.0m.pdf

60. The PANDA Experiment: http://www-panda.gsi.de/

61. The CBM Experiment: http://www-cbm.gsi.de/

62. Job statuses in AliEn: http://pcalimonitor.cern.ch/show?page=jobStatus. html

63. LHC Design Report: http://lhc.web.cern.ch/lhc/LHC-DesignReport.html

64. LHC Performance and Statistics: https://lhc-statistics.web.cern.ch/LHC-Statistics/

65. The ALICE Collaboration: First proton-proton collisions at the LHC as observed with the ALICE detector: measurement of the charged-particle pseudorapidity density at $\sqrt{s}$ = 900 GeV, Eur. Phys. J. C65 (2010), 111-125.

66. Quark Matter 2011: http://qm2011.in2p3.fr/

67. The ALICE Collaboration: Charged-Particle Multiplicity Density at Midrapidity in Central Pb-Pb Collisions at $\sqrt{s_{NN}}$ = 2.76 TeV, Phys. Rev. Lett. 105 (2010), 252301.

68. The ALICE Collaboration: Elliptic Flow of Charged Particles in Pb-Pb Collisions at $\sqrt{s_{NN}}$ = 2.76 TeV, Phys. Rev. Lett. 105 (2010), 252302.

69. Physics Publications of the ALICE Collaboration in Refereed Journals, http://aliceinfo.cern.ch/Documents/generalpublications

70. I. Foster et al: Cloud Computing and Grid Computing 360-Degree Compared, Proc. of the Grid Computing Environments Workshop, 2008. GCE'08, Austin, Texas, http://arxiv.org/ftp/arxiv/papers/0901/0901.0131. pdf

71. C. Loomis: StratusLab: Enhancing Grid Infrastructures with Cloud and Virtualization Technologies, TERENA Networking Conference (TNC2011),Prague,   2011,   https://tnc2011.terena.org/web/media/ archive/11C

72. LXCloud: https://twiki.cern.ch/twiki/bin/view/FIOgroup/LxCloud

73. The ATLAS Collaboration: Search for the Higgs boson in the H → WW → lvjj decay channel in pp collisions at $\sqrt{s}$ = 7 TeV with the ATLAS detector; arXiv:1109.3615v1, Sep. 2011.

74. The ATLAS Collaboration: Search for a Standard Model Higgs boson in the H → ZZ → llvv decay channel with the ATLAS detector; arXiv:1109.3357v1, Sep. 2011.

75. The CMS Collaboration: New CMS Higgs Search Results for the Lepton Photon 2011 Conference; http://cms.web.cern.ch/news/ new-cms-higgs-search-results-lepton-photon-2011-conference

76. The LHCb Collaboration: A search for time-integrated CP violation in D0 → h +h − decays and a measurement of the D0 production asymmetry; http://cdsweb.cern.ch/record/1349500/files/LHCb-CONF-2011-023. pdf The LHCb Collaboration: Measurement of the CP Violation Parameter AΓ in Two-Body Charm Decays; http://cdsweb.cern.ch/record/1370107/ files/LHCb-CONF-2011-046. pdf

Chapter 7

QUALITY CONTROL AND CHARACTERIZATION OF SCINTILLATING CRYSTALS FOR HIGH ENERGY PHYSICS AND MEDICAL APPLICATIONS

Daniele Rinaldi[1], Michel Lebeau[2], Nicola Paone[3], Lorenzo Scalise[3] and Paolo Pietroni (formerly with[3])

[1]Università Politecnica delle Marche/ Dipartimento di Scienze e Ingegneria della Materia, dell'Ambiente e Urbanistica, Switzerland, Italy

[2]European Organization for Nuclear Research (CERN)/Department PH

[3]Università Politecnica delle Marche/ Dipartimento di Ingegneria Industriale e Scienze Matematiche

INTRODUCTION

To the discovery and first use of scintillators are linked the names of W. Crookes, A. Becquerel and E. Rutherford. Since then phosphors have been used to materialise information generated by various scientific, medical and industrial apparata.

Phosphorescence, luminescence and scintillation are basically the same phenomenon, differing by the internal mechanisms involved and by their decreasing time scales. Energy carried by radiative phenomena is converted into light in the phosphor when excited electrons turn back to their equilibrium state by the release of photons in the visible (or near-visible) range. In the first applications, observation was only visual and result qualitative. By their growing demand in space and time resolution, applications themselves prompted the increased performance of the phosphors. The early X-ray radioscopic devices, with their slowly glowing zinc sulphide screens, gave way to faster and safer means of observation.

Thanks to the development of adapted technologies (invention of the photomultiplier tube by Curran and Baker in 1944), light could be converted into electric analog signal and ultimately into manageable data. At that stage, the major interest of the phenomenon i.e. the proportionality of the light output to the incident energy, could be fully exploited. Experimentation in nuclear and particle physics began using plastic and liquid scintillators of increasing light yield and fast response. The discovery of NaI(Tl) in 1949 (Hofstadter, 1949) owed Hofstadter a Nobel Prize. Despite its hygroscopy and radioactive dopant, NaI(Tl) reached mass production scale, and is still in wide use today. With the discovery of BGO ($Bi_4Ge_3O_{12}$) scintillation in 1973, M.Weber and R.Monchamp (Weber, M. & Monchamp, R., 1973) opened the era of fast, dense synthetic mineral crystalline scintillators at the industrial scale: by 1989 $1.4m^3$ (11 400 pieces) of BGO had been produced for the L3 experiment at CERN. This achievement initiated the steady supply for a growing medical imaging market. Thanks to focused progress in solid state physics, deeper understanding of the physical phenomenon led to light production levels competing with the best organic scintillators without their weak sides. The role of R&D collaborations and dedicated conference cycles was crucial in this progress (SCINT conferences since 1991, Crystal Clear Collaboration since 1992). Attempts were made to produce amorphous (glass) and even polycrystalline (ceramics) scintillators, to try and gain from available profitable mass production methods. Finally the many advantages of monocrystalline structure has turned to be the mainstream in scintillator development. After NaI(Tl) and CsI(Tl) followed a sequel of new, better performing crystals. BGO, CeF_3, BaF_2, $PbWO_4$, LuAP, LSO, LYSO paved the way of a continued progress in performance but also in quality and quantity, the early formulae mostly thanks to the growing scale of high energy physics instruments, the more recent sustained by the growing demand and important economic prospects of medical imaging equipment.

This new generation of materials is characterised by an excellent time resolution, with a steep rise and short persistence (no afterglow), a chemical structure that guarantees reproducible properties and reliable performance over time and a resistance to working conditions (no aging, radiation resistance). By choosing high Z chemicals, high density crystals can be synthesised, ensuring the tight containment of deposited energy -thus reducing the instrument dimensions. High purity raw materials, sophisticated production processes and adapted quality control methods ensure production of high grade crystals. Among several quality criteria the optical transparency in the scintillation wavelengths is a severe limiting factor as light attenuation and non-uniformity may deteriorate the crystal performance.

Not only intrinsic scintillators have been produced. Passive crystalline lattices have been designed to host specific chemical species –dopants- responsible for the light production, either by themselves or by their specific bonds with the lattice. This is a strong economic incentive as the host lattice may be made of less expensive materials, and budget better spent on costly dopants (e.g. rare earths). The dopant fraction is usually of a few percent. Dopants are selected to match the lattice properties to the best (crystal symmetry, lattice parameters). Two obstacles have to be overcome: segregation that makes light production uneven over the crystal volume, and lattice distortion that may induce mechanical stress and be detrimental to the production yield. Depending on the application and the selected scintillator, crystal sizes may be very different. From bulky 150cm^3 prisms used in high energy physics electromagnetic calorimeters, to mm^3 scale in medical imaging, production problems are quite different and may stress on different features for the optimisation of the material. The latter application has recently seen the development of crystalline fibres of the order of 1mm that may be the present industrial optimum at that scale.

In order to present methods for quality control of scintillating crystals, it is necessary to understand the production process and to identify the characteristic features which determine crystal performance and therefore will be subject to specifications that need to be verified. The following paragraph 2 describes crystal production process, paragraph 3 outlines applications of scintillating crystals, then paragraph 4 treats the methods for quality control based on photoelastic analysis and their applicability to process and product control of crystals. A final paragraph will resume the content of the chapter.

PRODUCTION PROCESS OF SCINTILLATING CRYSTALS

Scintillating crystal production has evolved from the chemical laboratory to the industrial plant scale. Crystals are produced by growth methods specific (optimal) of every chemical compound, part size and quantity (Lecoq, P. et al., 2006).

Raw Materials

Upstream the preparation of raw materials is a prerequisite of the ultimate crystal quality. Quality control and traceability have to secure a supply of tightly specified ingredients. Purity is not an absolute criterion but rather an economic compromise of innocuous and poisonous impurities, affecting scintillation (afterglow, light yield) transparency (colour centres) radiation resistance and built-in stress level (cell distortion). Stoichiometric proportions may not be the optimum as some components may be lost during the growth process, either by evaporation in the furnace atmosphere, or by combination with the crucible material.

Raw materials are ground to specific granularity distributions, thoroughly mixed to required proportions. Preparation is completed by melting the components and producing a polycrystalline compound that shall be used to fill the crucible for the growth operation.

Growth Methods

A variety of techniques is used to grow scintillating inorganic crystals. They are all derived from two main methods that shall be briefly described.

Czochralsky Method

In Czochralsky method (fig. 1), raw materials are molten in a metallic crucible and kept slightly above fusion point.

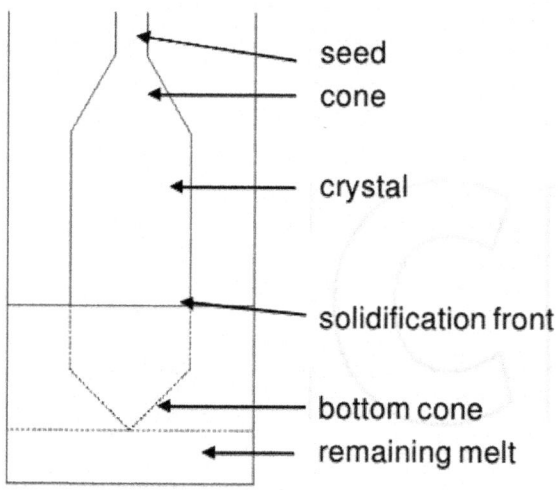

Figure 1: Czochralsky growth method (pull-from-melt).

A small monocrystal of the same material (seed) is put into contact with the molten bath and pulled up to lift a small meniscus of liquid by capillarity. Solidification occurs at a position and a rate fixed by several parameters. The thermal gradient is regulated by an induction loop heating the melt with help of the crucible mass. The melt temperature is homogenised by crucible regulation. The growth ("pulling") rate is monitored by a weighting system connected to the seed support. The rising solidified bulk takes the shape of the desired ingot (or boule). The pulling rate gives the ingot its conical top and bottom ends and its cylindrical overall shape. Optimal conditions result in turning the full melt into the ingot volume. The growth of a 1kg PWO ingot took of the order of 1 day. In spite of the circular symmetry of the whole system, the ingot shape may be facetted or oval (fig. 2), depending on the crystal lattice and on the seed orientation. The latter is a critical parameter.

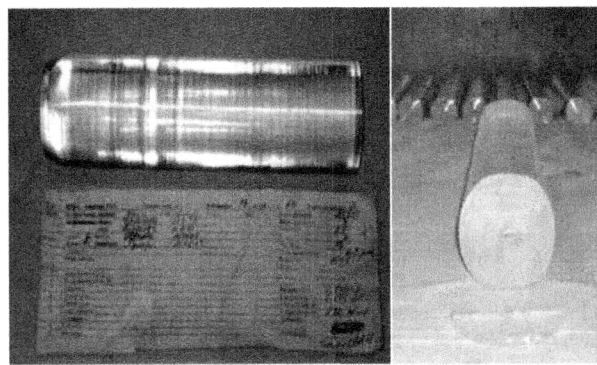

Figure 2: Czochralsky crystals. The ingot section results from growth parameters and lattice dimensions. Important material scrap is due to difference between ingot and piece, but "peel-like" scraps are recyclable.

The crucible is metallic to allow heating by induction. The metal is selected for a high melting point, and for a low chemical affinity with the melt. Platinum is the panacea but poses an economic problem because of its high cost and stock price volatility. The crucible raw material is therefore usually leased from a specialised bank. Even after a large production, the crucible material loss remains small and amounts for a minor fraction of the production cost. Some less expensive metals like palladium or osmium may be used depending on the crystal. The melt free surface is subject to chemical exchanges with the furnace atmosphere, with the risk of losing the design proportions. This is therefore a critical feature of the method and controlled atmosphere is the rule. Depending on the crystal material, vacuum, inert, reducing or oxidizing atmospheres are chosen. For instance, in the case of PWO growth, lack of oxygen induces mechanical stress, whereas its excess is detrimental to crystal transparency.

Bridgman Method

This method is an evolution of the fusion zone refining method (fig. 3). Work piece shape is given by grinding off, powder scrap is not recyclable. The polycrystalline material is shaped to the final part proportions with extra thickness for the mechanical processing to follow. The shape is usually an elongated prism. A monocrystal seed is placed at one end. This assembly is contained in a thin (a few tenths of a mm) metallic envelope, usually made of platinum. One or several of these assemblies are positioned in the furnace enclosure. The fusion zone (thermal gradient) moves from the seed interface along the ingot length by combining several induction loops. The gradient displacement velocity is comparable to the Czochralsky growth rate, but the operation has to be repeated several times (at the week scale) to reach monocrystallinity and purity at the required level. The Bridgman process competes with the Czochralsky method only by the use of multiple ingot furnaces. The last passes are considered as an annealing phase. After cooling down the metallic envelope is peeled off and recycled.

Figure 3: Bridgman crystals. The ingot shape is given by an envelope-like crucible.

Annealing

The solidification process results in thermally-induced stresses. They are supposed to have the conventional parabolic profile, with moderate compression in the core, and higher tension at the periphery. The latter is a matter of concern as crystals are especially prone to breaking in tension (crack opening). Annealing is necessary to reduce these tensions. The annealing temperature is very close to the melting point to ease dislocation movements and resolution. The thermal effect is sometimes combined with a chemical one with help of a controlled atmosphere, as a chemical correction is expected to restore the crystal lattice equilibrium. In that case annealing time is governed by diffusion and may take several days. The annealing furnace is therefore a complex device and annealing a costly operation in time and money.

Mechanical Processing

Ingots are the base material for expensive parts tightly specified in shape and dimensions. Mechanical processing is designed to achieve these two goals.

For calorimeters, ingots are usually dimensioned for the yield of one piece. Attempts were made to get two and even four pieces out of an ingot. The tiny pixels of medical imaging are either obtained by the many from a single ingot, or cut from fibres.

Cutting (Shaping)

The cutting operation provides the part its general shape. Precision is required to limit the amount of material to be removed later by lapping and polishing. It is therefore an economic target to obtain the best geometry (planarity, correct angles) at cutting. Processing parameters must be optimised to save on time, but not to the price of too thick a damaged sub-surface, and increased risk of edge chipping. The surface finish obtained at cutting ranges from 0,2 to 0,05 μm Ra. Finer finish is not worth doing at cutting as lapping easily produces values below 0,02 μm Ra. A very delicate feature of the cutting operation is the balancing of released thermally-induced tensions as the ingot periphery is removed, face by face. Operations have to be sequenced in such a way that this transitory unbalance is not aggravated (fig. 4).

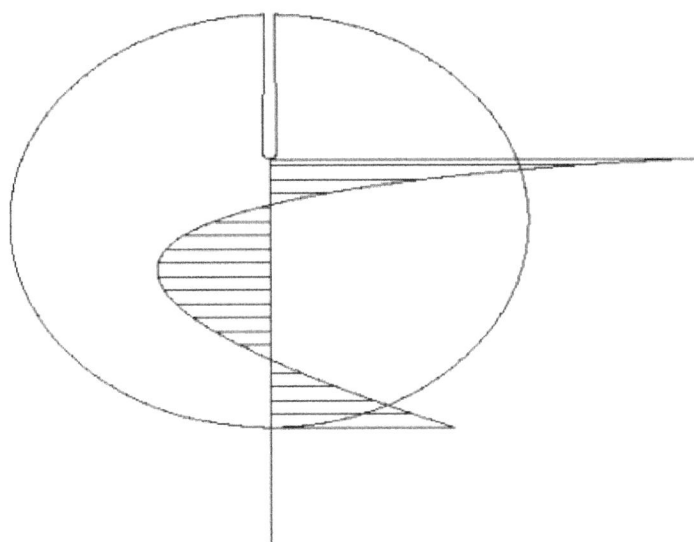

Figure 4: Effect of cutting on tensional state balance; Cutting a tensioned crystal "frees" mechanical tensions: New boundary conditions transform mechanical tensions into material displacements and deformations, and new mechanical tension distribution. Complex problem with risk of even higher concentrated stresses and breaking when cutting does not follow crystal lattice symmetry.

Figure 5: Crystal shaping by grinding off.

Thanks to their transparency, scintillating crystals may be subject to photoelastic stress observation. This method will be discussed in paragraph 4, and it was specifically applied to BGO (Rinaldi et al., 1997) and later to PWO (Cocozzella et al. 2001, Lebeau et al., 2005). The resulting information, as well as lattice orientation and mechanical properties are critical inputs for determining the processing conditions (Lebeau, 1985, Ishii & Kobayashi, 1996, Pietroni et al., 2005). Grinding-off is the oldest method, that consists in removing excess material by multiple passes of a grinding wheel (fig. 5), on a conventional planar grinder.

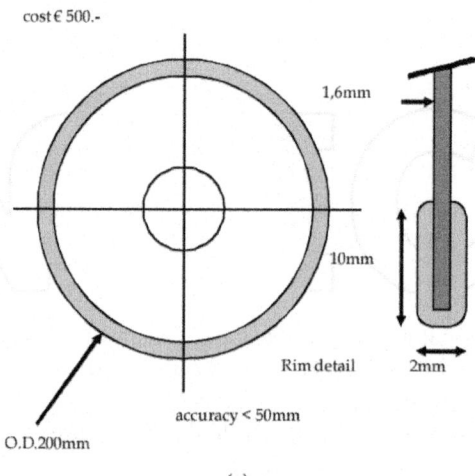

(a)

Disc edge showing lines left by truing (dressing)
Bright dots are diamond grains

1mm

76mm

2,2kts/cm³
or
550000 grains /cm³

average grain spacing
150mm

Detail of embedded diamond grains (nominal size 76µm)

0,25 mm

(b)

Figure 6: Cutting disc (a) and disc details (b).

This is a safe, very accurate but lengthy operation that has been replaced by more economic solutions. Thin disk saws with abrasive-loaded rim are now the most commonplace tool for crystal cutting (fig. 6).

The abrasive grain selection is capital. A good formula is 50 to 100 µm diamond grain sintered in high density in a bronze matrix. The disc diameter is selected to achieve a complete cut in one pass, e.g. 200mm for a 30mm cut depth (fig. 7).

(a)

Cutting disc mounted on machine spindle
(balancing head removed)

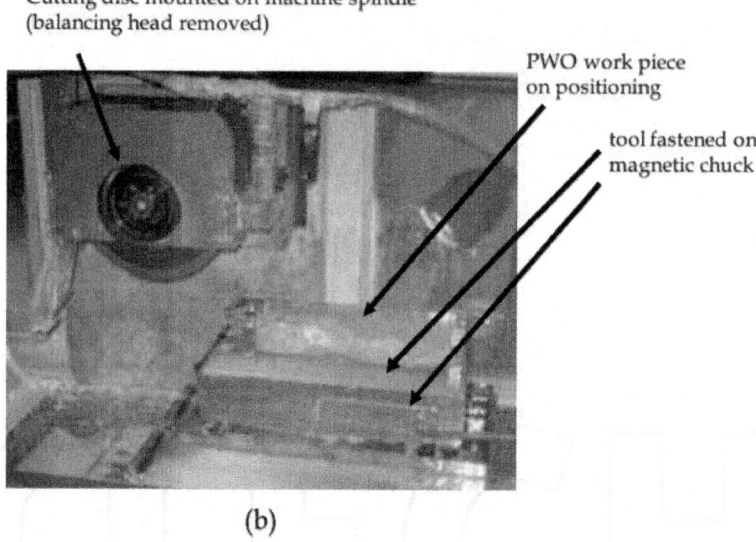

PWO work piece
on positioning

tool fastened on
magnetic chuck

(b)

Figure 7: Crystal cutting machine (a) adapted grinder, CERN, 2000, detail of cutting machine (b) Details of the cutting machine.

The best planarity achieved with this method is of the order of 20 μm, and depending on the tooling and machine quality, a dimensional precision of the order of 50 μm is expected. The disc thickness is a compromise between the cutting precision and the ingot material loss (as fall-off is recycled, but not cutting swarf). The disc thickness also contributes to subsurface damage depth. A thickness of 2mm is the optimum for a 200mm disc. To achieve the above-mentioned precision the best discs on the market have to be purchased. The machine tool is derived from a conventional grinder, with improved casing for intensive lubrication (fig. 8, a).

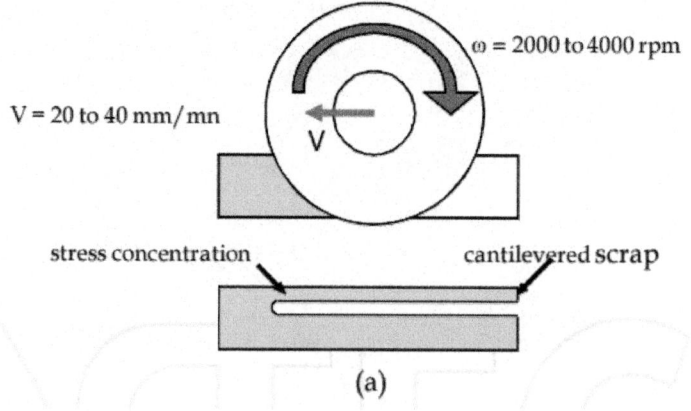

$\omega = 2000$ to 4000 rpm

$V = 20$ to 40 mm/mn

V

stress concentration

cantilevered scrap

(a)

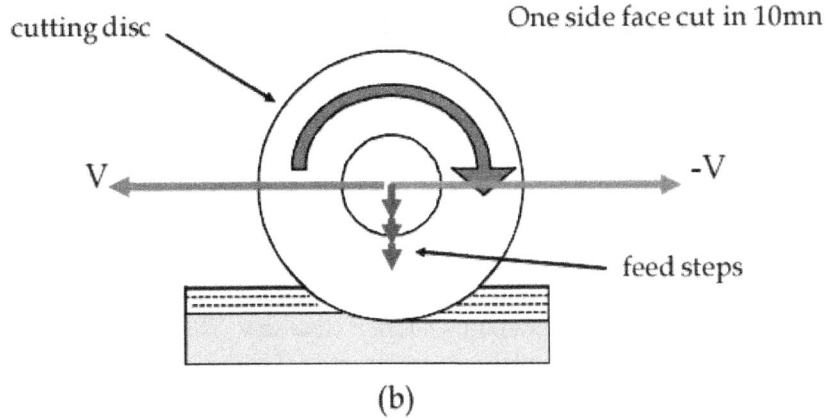

(b)

Figure 8: Full-depth cutting (a); sweeping mode (b) keeps some stress balance.

The disc velocity is about 3000rpm, the feed speed about 40mm/min. Sweeping mode: instead of cutting one crystal face in one stroke of the cutting disc, several passes at higher feed (meter per second) with a small vertical take (0,1mm) at every pass are combined for a similar processing time (fig. 8, b). The main advantage is a smaller pressure in the cutting zone, and a balanced release of tensions as the work piece remains symmetrical during the operation. The internal disc saw method was widely used for wafer cutting before the generalisation of plain wire cutting. A very thin (0,3mm) metallic circular membrane tensioned on its periphery is tensioned to achieve a stable planar shape. The edge of the central circular opening is bordered with sintered abrasive grains (fig. 9).

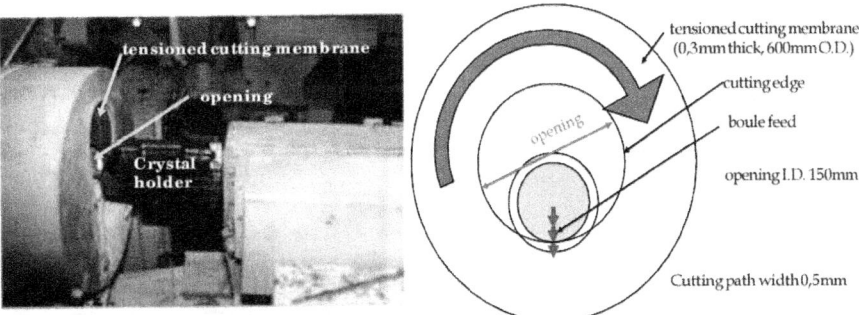

Figure 9: Inner cutting machine.

Only bars of small section compared to the disc opening can be cut with this method. The advantage is an economy in material thanks to the narrow cutting path (disc thickness plus 0,1mm max.). Processing of prismatic pieces is only possible for end cuts. Because of the limited free space within the central opening supporting tooling design is critical.

The plain wire saw method, as said above, is now widely used for mass production of silicon wafers, for electronics as for solar cells. The km long wire runs back and forth and follows a complex path to achieve multiple cutting planes on several ingots (today up to seven with diameters exceeding 320mm). The abrasive slurry (usually cheap corundum) is poured on the wire where it holds by capillarity. The cutting action depends on the grain adherence to the wire, with the result of decreasing efficiency with cut depth. Wire diameter and cutting path are comparable to internal disc saw. This method therefore requires a correction of the planarity afterwards. The specific arrangement of this equipment is only fit for slicing and has no interest for the shaping of prismatic scintillators. Wires with sintered abrasive have been successfully developed to correct the weak sides of the plain wire. Typical diameter is 0,25mm with an 80 μm diamond grain coating. The 2km wire is expensive (1 €/m order) and fragile: processing parameters and lubrication have to be carefully adapted to dedicated machine tools (fig. 10).

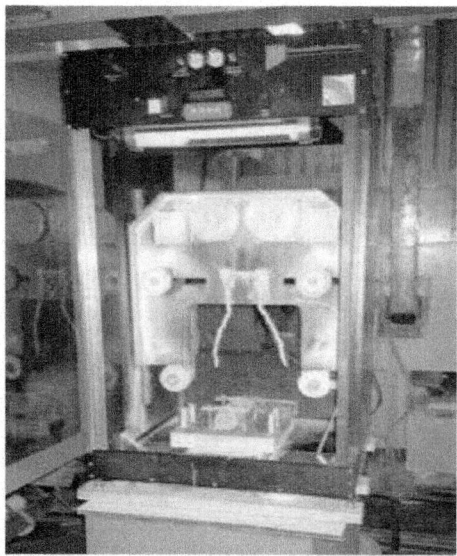

Wire length 2km @ 2€/m
Wire O.D. 0,25mm
Diamond grain 80 Hm
Wire path 0,3mm
Wire speed 5 to 10m/s
Feed 25 to 50Hm/s

Figure 10: Wire saw (abrasive wire).

Feeds of 50 μm per minute can be achieved with an excellent planarity and a very low subsurface damage. Parameter optimisation also aims at reducing the wire wear. By combining the feed with the crystal rotation, a symmetric end-cut is possible (cropping), with a balanced stress relief (fig. 11). The machine open configuration allows cutting long side faces (up to 300mm).

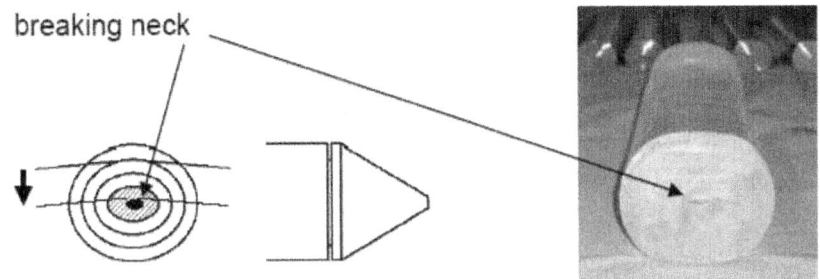

Figure 11: Rotary wire end-cut (solves boule-ends tensions release); Large ingots have to be put to length before annealing because of annealing furnace dimensions. Cutting un-cured ingots is very delicate and a rotary method is used to keep some symmetry. The cutting wire is slowly fed down while ingot rotates until the end breaks at the thin remaining neck.

Crystal cutting is an abrasive process at the microscopic scale. Every abrasive grain works as a gross tool with a negative cutting angle that locally induces high compressive stress. To prevent high crack density and possible propagation, reduce tangential forces, keep work piece temperature low and ease chip removal, the appropriate lubricant must be applied in abundant flow. pH, chemical polarity and affinity may be adapted to the crystal material in a profitable way. Filtering, sedimentation and recycling are environmental constraints. Lapping is free abrasive action between the crystal face and the surface of a rotary table, the lap (fig. 12).

Combined rotations of the lap and the crystal result in an even distribution of the abrasive action and a regular material removal. Working parameters are the lap and crystal rotation velocities (a few m/s), the pressure exerted on the crystal face (a few N/cm^2), the abrasive material and granularity (usually about 15 μm corundum or diamond), the lubricant mixed with the abrasive (slurry) , and finally the lap material. A typical stock removal for PWO was 50 μm/min. With a 0,02 μm Ra finish reached after 3 min, the damaged subsurface layer from cutting was easily removed. This finish Ra is a good value to start polishing. To prevent edge chipping and resulting deep scratches on the surface, chamfers are necessary on every sharp edge of the crystal before lapping (and polishing): 0,2-0,3mm bevels are usually sufficient. Polishing produces optically transparent faces, that are necessary for scintillating light collection (Auffray et al. 2002). The polish quality can be specified according to a maximum number of visible scratches per view field at a given magnification. The value is far less demanding than for conventional lens polishing. Scintillator polishing operates in similar configuration as lapping. The main differences are the abrasive grain size (from 3 down to fraction of a μm), and the lap cover. Because of the abrasive fine grain, stock removal is slow (less than 1 μm/min) and polishing takes 10 to 20 min per face. This is the critical path in a crystal processing line (Auffra et al. 2002). In mechanical polishing, the material removal results from grain abrasion as for lapping, but at a smaller scale (fig.13). Diamond is the best abrasive in that case. Cooling and lubrication are critical to avoid subsurface damage.

This method was developed for electronic chips and finds interesting developments for scintillators. The abrasive action is enhanced by specific chemical conditions. For instance, a suspension of very fine grains of quartz (20 nm) in pH9 colloidal silica produces an efficient polishing free of sub-surface damage (Mengucci et al., 2005). Soda and potash were also tested but surface etching sometimes happened when mechanical action was not properly balanced. The use of aggressive chemicals (bases) poses difficult safety and environment conditions that prevent the spread of these methods. Cerium oxide is known for its combined abrasive action and chemical reaction with conventional lens glass. It is less delicate in use and has been successfully tested with PWO and LSO, but its chemical action remains unclear to date.

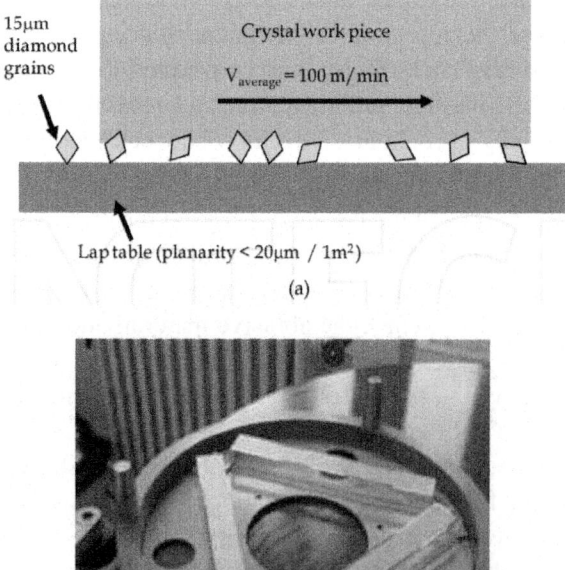

(a)

(b)

Figure 12: Lapping principle (a): Abrasive grains are bumped and tilted between lap and work piece and present fresh cutting edges to work. Lapping tooling (PWO,

CERN, 2000) (b): Three crystal shapes are cut out in the lapping mask (or holder). A satellite ring keeps the mask (and crystals inside) in radial position on the lap. Crystal length 230mm, ring I.D. 320mm.

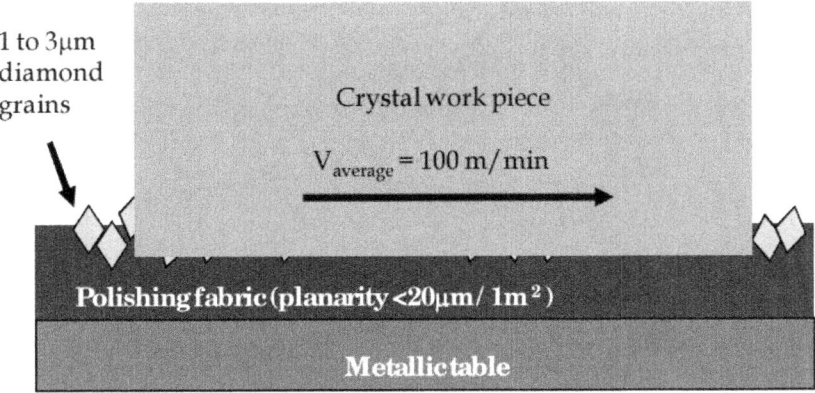

Figure 13: Polishing schematics. Diamond grains are taken in stable cutting orbits by the fabric. Material removal operates at microscopic scale and some ductile effect results, with limited subsurface damage.

SCINTILLATING CRYSTALS: APPLICATIONS FIELDS

In recent years, scintillating crystals have found numerous applications in different fields; hereafter we will briefly recall the main areas: Nuclear and high energy physics, medicine (imaging of biological tissues), geology and security.

Nuclear and High Energy Physics

This domain is where scintillators were discovered and where most of their development took place. The early 20th century saw their use in fundamental research. The atomic era opened by the 2nd World War multiplied the use of scintillating counters, also necessary in nuclear energy production. The quick development of fundamental research in high energy and particle physics after the war was a stimulating motivation for increased performance, quantity and economy. The most striking example is in electromagnetic calorimeters (fig. 14), with new projects involving tons of the most recent scintillators (LYSO). Medical imaging benefits of the spin-off of this striving discipline.

PbWO$_4$ electromagnetic calorimeter

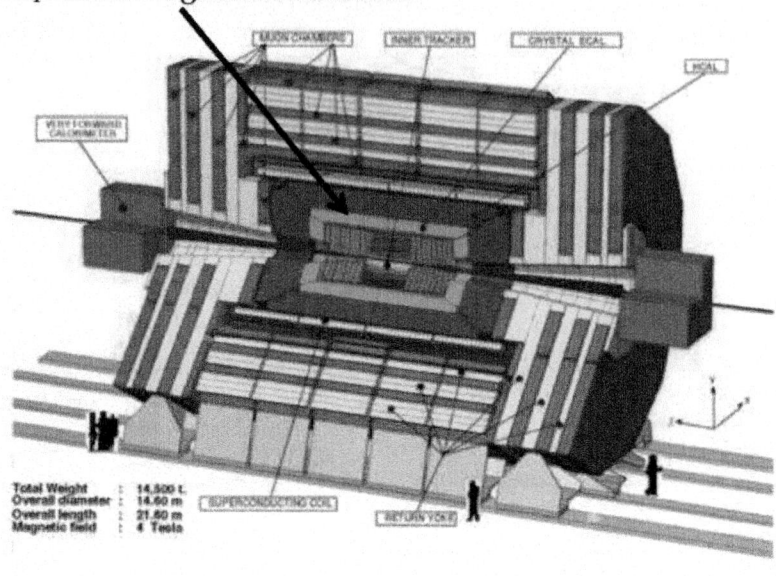

(a)

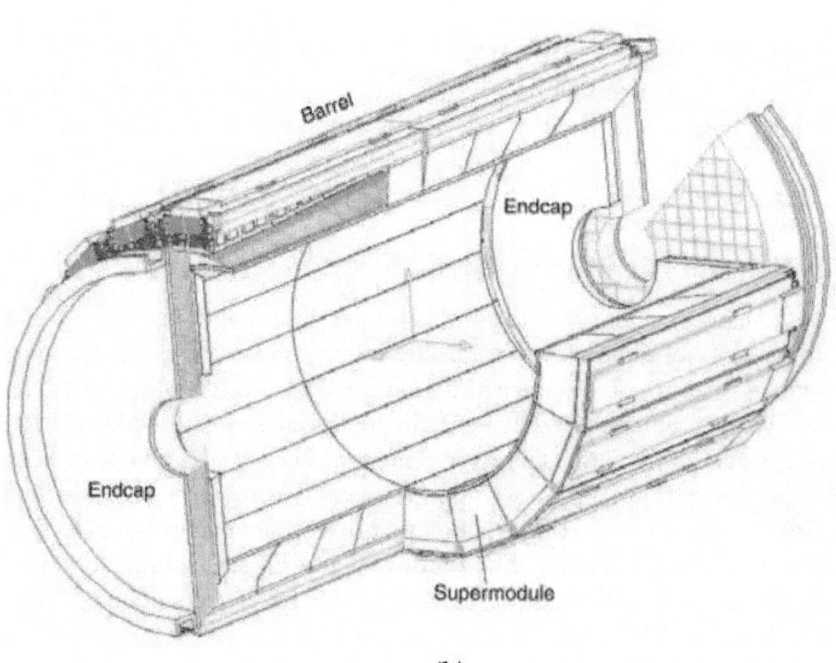

(b)

(c)

Figure 14: The CMS experiment at CERN LHC with PbWO4 electromagnetic calorimeter (a), the CMS PbWO4 electromagnetic calorimeter Modularity crystal sub-module of 10 crystals module of 400 (500) sub-modules super-module of 4 modules barrel of 36 super-modules (b) and (c) a module of the CMS PbWO4 electromagnetic calorimeter (10 x 20 crystals arranged in pointing geometry).

Medical Imaging

Modern radiography is characterised by lower radiation doses (and/or shorter exposures), 3-D information (tomography), real-time observation, tissue or function identification, with the help of large arrays of fine scintillators (pixels) surrounding the patient or covering the organ (mammography) and powerful reconstruction software. The two main types are projective imaging (e.g. X-ray CT) and PET scanners. In X-ray CT radiation scans the patient from outside and projective information is reconstructed. In PET the patient ingests some radio-element that releases positrons (beta decay). The recombination of the positron with an ambient electron produces two opposite gamma rays of well defined energy detected in opposite scintillator pixels. The radio-element is combined in a chemical (tracer) specific of of tiny 2 x 2 x 10 mm^3 prisms. Processing was dominated by material loss and sub-surface damage because of the crystal scale compared to processing tools. Scintillating fibres are an interesting solution. New fast scintillators with high light output like LuAP, LSO and LYSO are very promising for the medical domain, where economic prospects are high.

Geologic Research

Mining, gas and oil logging are very active economic domains, because of the increasing demand in raw materials and fuel, opposed to the shortage forecast and resulting crisis. Research is performed by prospective drilling. The drill hole may be scanned with two types of detectors. The simpler one is a radiation detector used for the research of radioactive minerals. B.Pontecorvo already proposed such a device with a ionizing chamber counter in 1941. The other type contains a powerful neutron source (Cf, Am, Be, Cs) that irradiates the underground vicinity of the hole. Stimulated gamma emission reaching the detector is typical of the chemical bonds in the mineral (hydrocarbon). Obviously the detector is shielded from the source. The energy spectrum typical of the concerned mineral, and the signal intensity may give a quantitative information. The detector transmits its signal to the recording station on the surface. The source is usually left in the drill hole bottom for safety reasons. Today detectors are of the scintillator type (NaI(Tl), BGO, GSO, LuYAP). Literature is scarce because of patent protection.

Security

Quick, non-invasive inspection of transport loads, containers, but also luggage and passengers is familiar to everybody in today's life. In the latter case, soft X-ray scanners are used to reveal hidden weapons or hazardous objects, thanks to their density or specific form. Nuclear explosives are also detected by portable radiation detectors. Scintillators are used in these applications. Dual (or multiple) energy systems combined with colour coding display help material identification (different Z) against apparent density. Visual training for qualitative identification are crucial to the efficiency of these detectors. Scientific literature lacks because of patent protection

Photoelastic Methods for the Quality Control on Scintillating

Paragraph 2 has presented the various phases of crystal production process. It emerges that internal stress distribution is influenced by the mechanical and thermal processes that the crystal undergoes. This brings to the issue of quality control, which implies the assessment of internal stress state in order to set up a proper production process and during its operation.

Residual stress is indeed a major hazard in crystal processing. Crystals are brittle materials, therefore residual tensile stress may easily lead to fracture and breakage during processing or, even worse, during the following assembly of many crystals into complex geometry detectors. Scintillating materials are transparent and usually are optically anisotropic. Internal stress causes lattice strain and deformations, which manifest as stress induced birefringence; this means that the piezo-optic properties of the material can be observed to verify its internal strain (or stress) state. Photoelasticity is a classical measurement technique suited to observe stress induced birefringence in transparent materials (Wood E., 1964, Dally J. & Riley W. 1987). Therefore photoelasticity is a natural candidate method for quality control of scintillating crystals; of course this measurement method provides only information on quality related to mechanical strain and stress, and requires an accurate knowledge of piezo-optic properties of the material, which is not always available. Furthermore, it is a volumetric technique which provides information on the spatial integral of the stress distribution along the light path through the crystal and not local values.

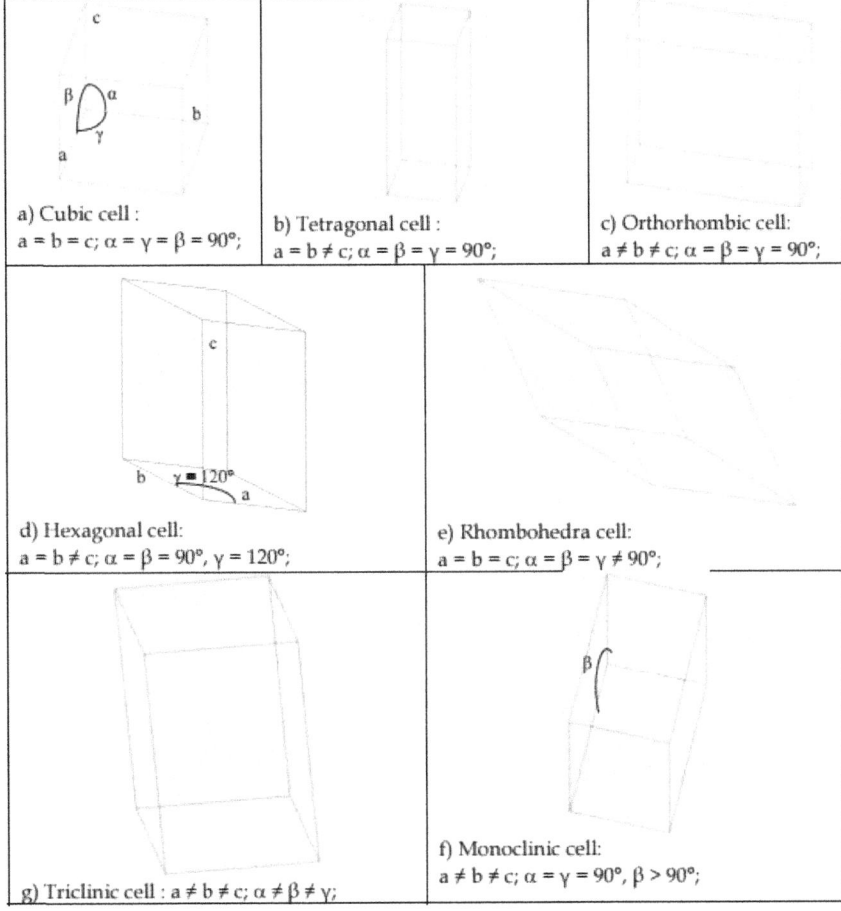

a) Cubic cell :
$a = b = c; \alpha = \gamma = \beta = 90°;$

b) Tetragonal cell :
$a = b \neq c; \alpha = \beta = \gamma = 90°;$

c) Orthorhombic cell:
$a \neq b \neq c; \alpha = \beta = \gamma = 90°;$

d) Hexagonal cell:
$a = b \neq c; \alpha = \beta = 90°, \gamma = 120°;$

e) Rhombohedra cell:
$a = b = c; \alpha = \beta = \gamma \neq 90°;$

g) Triclinic cell : $a \neq b \neq c; \alpha \neq \beta \neq \gamma;$

f) Monoclinic cell:
$a \neq b \neq c; \alpha = \gamma = 90°, \beta > 90°;$

Figure 15: Geometrical shapes of crystals.

Hereafter it is discussed crystal optics, optical anisotropy, piezo-optic behaviour and then photoelasticity is presented for crystal quality control.

Geometric, Mechanical and Optical Proprieties of Crystals

Crystal Lattice and Symmetry

A crystal is a solid material constituted by a 3D ordered structure which has the name of crystal lattice. Each crystal lattice is formed by the repetition of a fundamental element, the primitive unit cell: thanks to its replication, it produces the crystal structure (Wood E., 1964, Hodgkinson W., 1997, Wooster W., 1938.). From a geometrical point of view, it is possible to build up the crystal lattice simply translating the unit cell in parallel way with respect to its faces. Indeed the cell geometry should have peculiar characteristics: in particular, the opposite faces should be parallel and, for this reason, it should be a parallelepiped. Possible geometrical shapes are hereafter reported. The crystal physical and optical properties depend on the typology of unit cell and on the atomic bondages strength. Indeed those properties have the same symmetries of the crystal structure.

Elastic Properties of Crystals

Crystals undergoing a mechanical stress will deform, so they will exhibit an internal strain distribution. If the mechanical stress is below a limit, named elastic limit, crystal deformation is reversible. The strain is proportional with the applied stress for low level stresses. If the crystal undergoes an arbitrary uniform stress $[\sigma_{\kappa\lambda}]$ the generated strain components ε_{ij} is linearly correlated with the stress tensor (Wood E., 1964). This means that:

$$\varepsilon_{ij} = s_{ijkl}\,\sigma_{kl} \quad (i, j, k, l = 1, 2, 3)$$

(1)

Equation 1 is the generalized Hook law. Here, s_{ijkl} factors are crystal elastic compliances. The total number of the elastic compliances s_{ijkl} is 81. The Hook law can be written in the following way:

$$\sigma_{ij} = c_{ijkl}\,\varepsilon_{kl} \quad (i, j, h, l = 1, 2, 3)$$

(2)

Where c_{ijkl} are crystal elastic stiffness coefficients. The coefficients c_{ijkl} and s_{ijkl} form a forth order tensor. This means that in a coordinate system transformation from a coordinate system X_1, X_2, X_3 to X'_1, X'_2, X'_3 the coefficients s_{ijkl} (c_{ijkl}) are transformed into s'_{mnop} (c'_{mnop}) throughout the law:

$$s'_{mnop} = C_{mi}\,C_{nj}\,C_{ok}\,C_{pl}\,s_{ijkl}$$

(3)

where C_{mi}, C_{nj}, C_{ok}, C_{pl} are direction cosine which define the X_1,X_2,X_3 axes orientation with respect to X'_1,X'_2,X'_3. axes. Each s_{ijkl} (c_{ijkl}) coefficient has a precise amplitude and correlation with respect to a specific coordinate system, linked to the crystal. If this coordinate system is coincident with the crystallographic one, the coefficients are the basic ones. Since the strain and stress tensors are symmetrical, the tensor coefficients c_{ijkl} and s_{ijkl} are symmetrically coupled according to the subscript i and j, k and l, so:

$$s_{ijkl} = s_{jikl,} \quad c_{ijkl} = c_{jikl,} \tag{4}$$

$$s_{ijkl} = s_{jilk,} \quad c_{ijlk} = c_{ijkl,} \tag{5}$$

The equations (4) and (5) reduce the number of independent components of c_{ijkl} and s_{ijkl} to 36. Since c_{ijkl} and s_{ijkl} are symmetrical with respect to the first two subscripts and the second ones, the equations (4) and (5) can be written in more compact way:

$$\left.\begin{array}{l} s_{ij} \ (i,j = 1,\ 2,\ 3,\ 4,\ 5,\ 6) \\ c_{ij} \ (i,j = 1,\ 2,\ 3,\ 4,\ 5,\ 6) \end{array}\right\} \tag{6}$$

This notation reduces the number of terms of (1) and (2)

$$\left.\begin{array}{l} e_i = s_{ij}s_j \ (i,j = 1,\ 2,\ 3,\ 4,\ 5,\ 6) \\ s_i = c_{ij}e_j \ (i,j = 1,\ 2,\ 3,\ 4,\ 5,\ 6) \end{array}\right\} \tag{7}$$

but the following rules should be respected:

$$\left.\begin{array}{l} s_{ijkl} = s_{mn} \text{ when m and n are equal to 1, 2, or 3} \\ 2s_{ijkl} = s_{mn} \text{ when m or n are equal to 4, 5, or 6} \\ 4s_{ijkl} = s_{mn} \text{ when m and n are equal to 4, 5, or 6} \end{array}\right\} \tag{8}$$

It is necessary to underline that the symmetry further reduces the number of independent coefficients c_{ij} and s_{ij}. The following formula relates the elastic compliances s_{ij} to the elastic stiffness c_{ij}:

$$s_{ij} = \frac{(-1)^{i+j}\Delta c_{ij}}{\Delta^c} \tag{9}$$

Where Δc is a determinant composed of elastic stiffness:

$$\begin{vmatrix} c_{11} & c_{12} & c_{13} & c_{14} & c_{15} & c_{16} \\ c_{12} & c_{22} & c_{23} & c_{24} & c_{25} & c_{26} \\ c_{13} & c_{23} & c_{33} & c_{34} & c_{35} & c_{36} \\ c_{14} & c_{24} & c_{34} & c_{44} & c_{45} & c_{46} \\ c_{15} & c_{25} & c_{35} & c_{45} & c_{55} & c_{56} \\ c_{16} & c_{26} & c_{36} & c_{46} & c_{56} & c_{66} \end{vmatrix}$$

and Δc_{ij} is the minor obtained from this determinant by crossing out the i-th row and j-th column. Likewise:

$$c_{ij} = \frac{(-1)^{i+j} \Delta s_{ij}}{\Delta^s} \tag{10}$$

The following constants are often used for a description of elastic properties of both isotropic and anisotropic media. Young's modulus E, characterizing elastic properties of a medium in a specific direction, is defined as the ratio of the mechanical stress in this direction to the strain it produces in the same direction. The Poisson ratio v is defined as the ratio of the transverse compression strain to the longitudinal tensile strain caused by a mechanical stress. The Shear modulus μ is defined as the ratio of shear stress and shear strain, it produces in a material. In isotropic bodies only two of the above-mentioned constants are independent. For this reason, elastic properties of isotropic bodies are often described using the constants λ and μ, called the Lame constants. The constant λ and μ are related to the stiffness matrix components as follows:

$$\mu = \frac{c_{11} - c_{12}}{2} \tag{11}$$

$$\lambda = c_{12} \tag{12}$$

Considering the matrix (s_{ij}) then it is possible to write the following formulas E and μ in isotropic case:

$s_{11} = 1/E, \quad s_{12} = -v/E$ \hfill (13)

$2(s_{11} - s_{12}) = 1/\mu$ \hfill (14)

In anisotropic medium Young's modulus in a arbitrary direction X'_3 is:

$E = 1/s'_{3333}$ \hfill (15)

where $s'_{3333} = C_{3i} C_{3j} C_{3k} C_{3l} s_{ijkl}$ and $C_{3i}, C_{3j}, C_{3k}, C_{3l}$ are the direction cosine of the axis X'_3 with respect to the crystallographic coordinate system and s_{ijkl} are the basic compliances referred to crystallo-physical coordinate system. Young's modulus is a function of direction for all crystallographic classes, including cubic class.

In anisotropic media the Poisson ratio is equal to

$$v^{hk} = \frac{s_{hk}}{s_{kk}}$$

(16)

and it represents an estimation of lateral compression parallel to X_h with respect to accompanied elongation parallel to X_h.

Piezo-Optical Properties of Crystals

The piezoptical effect consists of changes in the optical properties of crystals throughout static and alternating external mechanical stresses and it is described in terms of the index ellipsoid. The general equation of the index ellipsoid in an arbitrary coordinate system X1, X2, X3, whose origin coincides with that of the main (crystallophysical) coordinate system, can be written in the following form:

$$B_{11}x_1^2 + B_{22}x_2^2 + B_{33}x_3^2 + 2B_{23}x_2x_3 + 2B_{13}x_1x_3 + 2B_{12}x_1x_2 = 1$$

(17)

Where B_{ij} are the dielectric impermeabilities or polarization constants. Equation 17 is related to anisotropic crystal without any applied mechanical stress.

An applied mechanical stress produces variation ΔB_{ij} in the dielectric impermeabilities:

$$\Delta B_{ij} = B_{ij} - B_{ij}^0$$

(18)

Considering a first-order approximation, the increments in the dielectric impermeability tensor components are proportional to mechanical stresses:

$$\Delta B_{ij} = \pi_{ijkl}\sigma_{kl}$$

(19)

On the other hand, the same increments can be expressed in terms of strain:

$$\Delta B_{ij} = p_{ijkl}\varepsilon_{kl}$$

(20)

Such a change in the optical index ellipsoid of the crystal due to the straining is called the elasto-optical effect. The coefficients π_{ijkl} and p_{ijkl} form a rank four tensor and they are called the piezo-optical and elasto-optical constants, respectively. In the matrix notation eqs. (19) and (20) can be rewritten in the following form

$$\Delta B_m = \pi_{mn} \sigma_n \tag{21}$$

$$\Delta B_m = P_{mn} \varepsilon_n \tag{22}$$

where if n = 1, 2, or 3:

$$\pi_{mn} = \pi_{ijkl}$$

while if n= 4, 5, or 6:

$$\pi_{mn} = 2 \cdot \pi_{ijkl}$$

and P_{mn} are the elasto-optical coefficients, $P_{mn} = p_{ijkl}$ for all m and n. In the general case:

$$\pi_{mn} \neq \pi_{nm} \quad P_{mn} \neq P_{nm}$$

The piezo-optic and elasto-optic coefficients are related by the following formulas:

$$P_{mn} = \pi_{mr} c_{rn}, \quad \pi_{mn} = P_{mr} s_{rn}$$

Where c_{rn} are the elastic stiffness and s_{rn} are the elastic compliances.

Isotropy and Anisotropy in Crystal Optical Properties

As it is described in previous paragraphs, the symmetry on the crystal lattice influences the symmetry on the optical properties. The Isotropy and anisotropy affect changes on refraction index and the consequent variations of the light velocity with respect to the direction inside the crystal material. Crystals, having a cubic cell, can be considered isotropic for their optical properties. All the rest of crystals cells has an anisotropic behaviour in terms of optical properties (Wood, E. 1964, Hodgkinson I., 1997, Wooster W., 1938). The optical anisotropy allows to classify crystals in two categories: uni-axial anisotropic crystals and biaxial ones according to the index ellipsoid which provides the value of the refraction index along a specified direction in the crystal (Wood, E. 1964, Hodgkinson I., 1997). A fundamental representation of Uniaxial anisotropic crystals is the optical indicatrix or ellipsoid of the refraction indices. As far as the uniaxial crystal, there are two principal indices: ordinary index of refraction n_o and extraordinary index of refraction n_e. Indeed the optical indicatrix is a rotation ellipsoid, where the both the axes are proportional to n_o and n_e. It is possible to state that an indicatrix is positive, when $n_e > n_o$, and negative, when $n_e < n_o$.

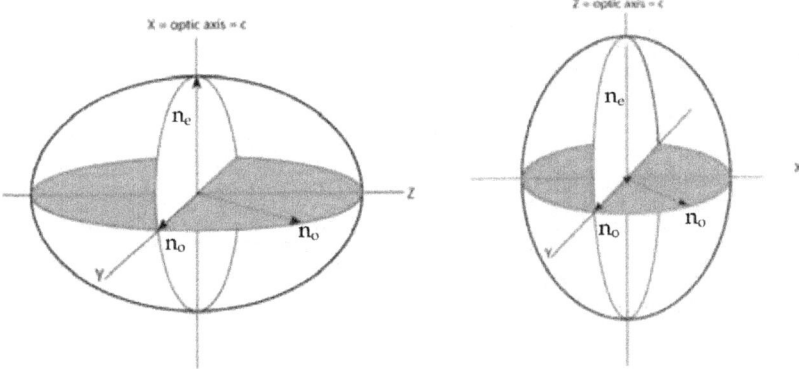

Figure 16: Optical indicatrix for uniaxial positive and negative crystal.

Hereafter three examples (fig. 17-19) are reported in order to explain the concept of uni-axial anisotropy and consequently of birefringence, assuming to study a positive crystal. In first case (fig. 17), the crystal lattice is oriented so that the optic axis is along the light travelling direction. In this case, considering the indicatrix its section perpendicular to the wave vector is circular with radius is n_o. Since in all the vibration direction of the electromagnetic field the refraction index is constant the birefringence is zero.

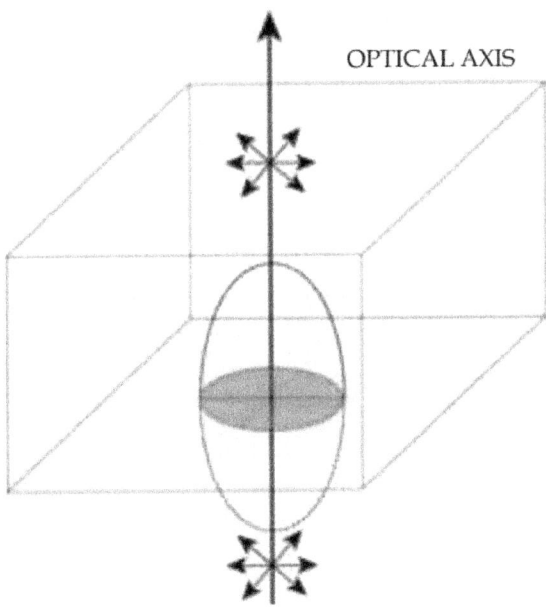

Figure 17: Uniaxial crystal first example.

In the second example (fig. 18), the crystal lattice is oriented in a random orientation so that the light path is at angle θ to the optic axis. The section through the indicatrix parallel to the incoming light wave is an ellipse whose axes are n_o and n_e. The extraordinary ray electromagnetic field vibrates parallel to the trace of the optic axis as seen from Fig. 18, while the ordinary ray one vibrates at right angles.

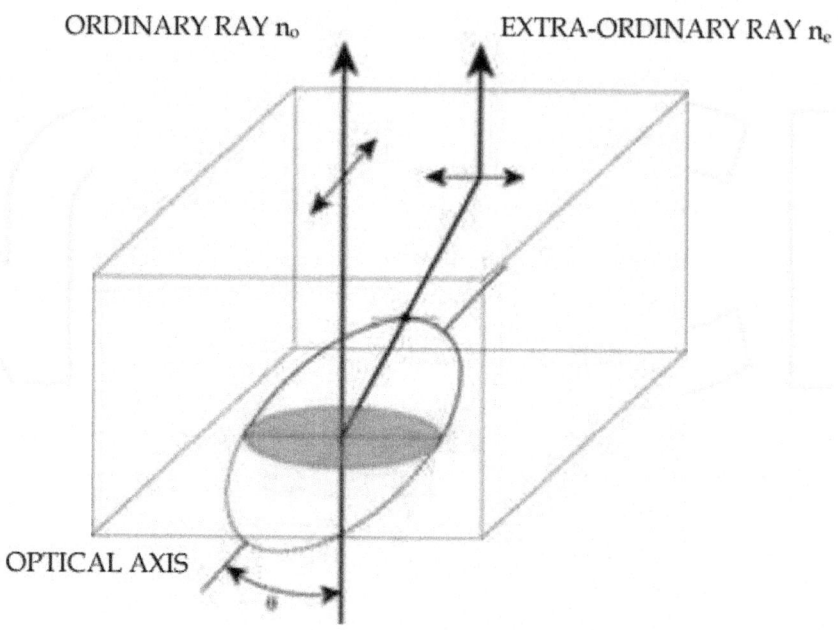

ORDINARY RAY n_o EXTRA-ORDINARY RAY n_e

OPTICAL AXIS

Figure 18: Uniaxial crystal second example.

In a third case (fig. 19), the crystal lattice is oriented so that its optic axis is parallel to the light wavefront. Because the optic axis has this orientation, this section is a principal elliptic section whose axes are n_o and n_e. The ordinary ray therefore has index of refraction n_o and the extraordinary ray n_e, which is its maximum because the crystal is optically positive. The extraordinary ray vibrates parallel to the trace of the optic axis (c axis) and the ordinary ray vibrates at right angles.

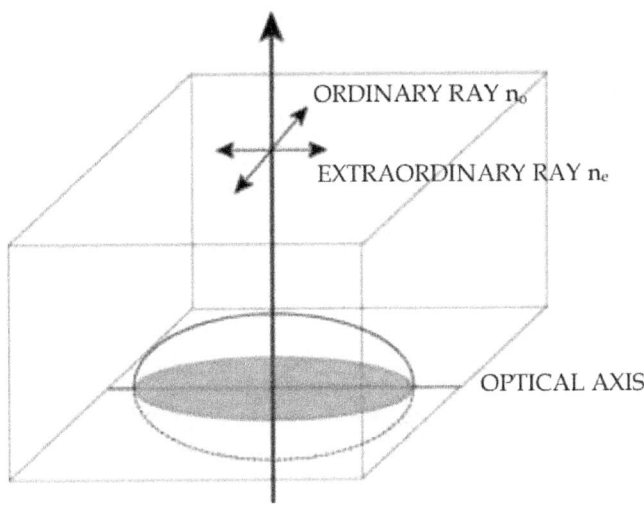

Figure 19: Uniaxial crystal third example.

These uniaxial crystals have tetragonal, rhombohedra and hexagonal cell.

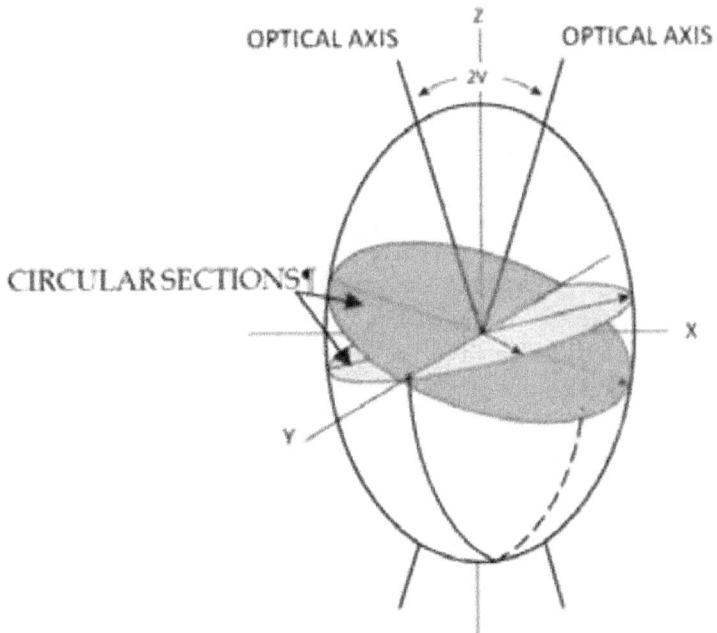

Figure 20: Optical indicatrix for biaxial crystals.

Those crystals having orthorhombic, monoclinic e triclinic cells are biaxial crystals. They have three different principal indices of refraction n_x, n_y and n_z, so that the indicatrix becomes a triaxial ellipsoid. Assuming that $n_x < n_y < n_z$, ZX plane is the optical plane and the Y axis is the optical normal (see fig. 20). Between the two optical axes an acute angle, 2V, is the optical angle. The bisector of such angle is the acute bisector Bxa (see fig. 21): in the positive crystal that is Z axis, while in negative ones that is the X axis. The bisector of the other obtuse angle between the optical axes is the obtuse bisector Bxo.

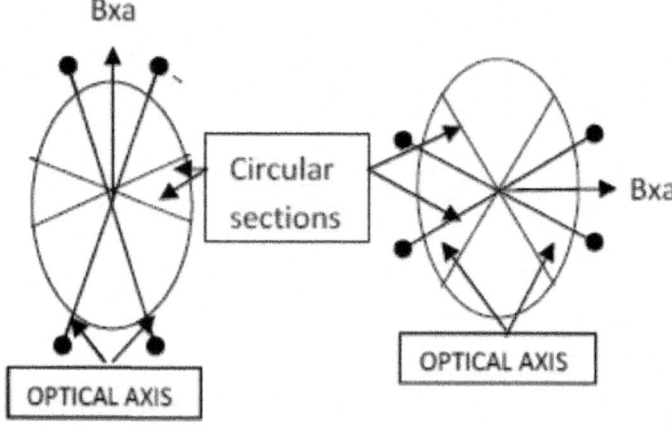

Figure 21: Optical indicatrix for biaxial positive and negative crystals.

The study of birefringence in this type of crystal can be conducted as in the case of uniaxial crystals. The indicatrix has the property that the axial sections normal to the optical axes are circular with a radius equal to n_y: then, a wave that propagates along the optical axis will behave as if they were moving in an isotropic medium. Any other section is elliptical wave moving along a direction different from the optical axis, therefore, will split into two beams, with vibration directions parallel to the major and minor semi-axis of the ellipse. The optical angle can be determined experimentally, but there are approximated formulas for its calculation according to the value of the indices of refraction:

$$tg^2V = \frac{\dfrac{1}{n_x^2} - \dfrac{1}{n_y^2}}{\dfrac{1}{n_y^2} - \dfrac{1}{n_z^2}}$$

(23)

The eq. (23) is valid for Z axis bisector: if $V > 45°$ the crystal is negative, while if $V < 45°$ the crystal is positive.

Photoelasticity

Photoelasticity is a classical technique that allows to visualize internal stress/ strain states in transparent materials; it exploits the changes in refractive indices induced by strain within transparent materials.

General Scheme of Polariscope

The polariscope is an optical instrument which utilizes polarized light in inspecting a specimen subject to strain; usually it is used to explore a two-dimensional planar stress state, with stress components orthogonal to the optical axis z. Light travels across the material of the specimen and its polarization state is affected by the spatial distribution of refraction index, which depends on strain. According to the kind of polarization, it is possible to consider a plane polariscope or a circular polariscope.

In a plane polariscope, devices known as plane or linear polarizer are utilized: they are optical elements which divide an incident electromagnetic wave in two components which are mutually perpendicular (fig. 22). The component, which is parallel to the polarization axis is transmitted, while the perpendicular one is absorbed or totally reflected internally.

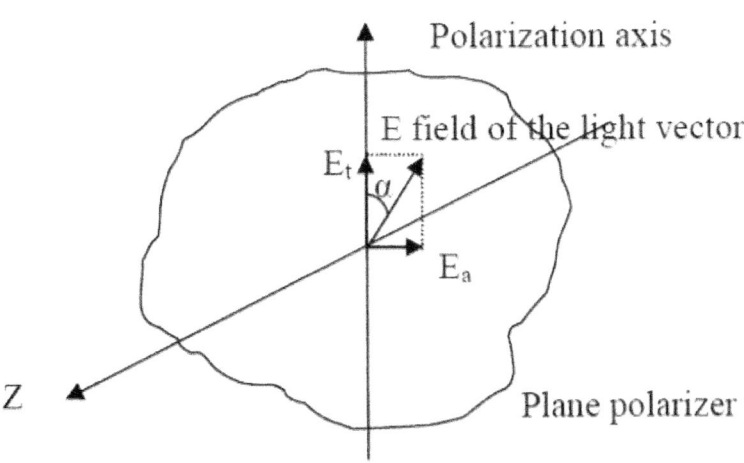

Figure 22: Plane polarizer.

It is possible to make the assumption that the polarizer is placed at the z_0 coordinate along the z axis, the equation of the light vector can be written:

$$E = a \cos \frac{2\pi}{\lambda}(z_0 - ct)$$

(24)

Since the initial phase is not important for this treatment, it is possible to rewrite it in the following way (it is assumed that $f = c/\lambda$ = wave frequency)

$$E = a \cos 2\pi ft = a \cos \omega t \qquad (25)$$

where $\omega = 2\pi f$ is the wave angular frequency. The absorbed and transmitted components of light vector are:

$$E_a = a \cos\omega \sin t \; \alpha \; E_t = a \; \omega t \cos \cos\alpha \qquad (26)$$

where α is the angle between the light vector and the polarization axis. In the plane polariscope two linear polarizers are used. Between those ones, the crystal under inspection is placed: the linear polarizer which is close to the light source is called the "polarizer", while that one placed on the opposite side with respect to the crystal is called the "analyser".

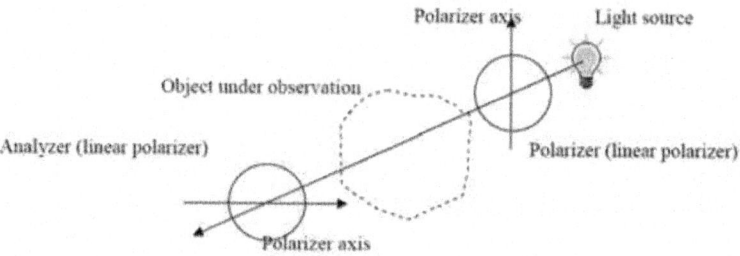

Figure 23: Plane polariscope scheme.

The usual configuration is the one where the two axes of polarization of analyser and polarizer are orthogonal to each other. The specimen to be analysed is put between the two, so that light goes through it. If the specimen is optically anisotropic, then light polarization is affected (see fig. 23). In our case the sample will be a crystal cut with plane surfaces. The advantage of such configuration is that what is observed is totally due to the crystal lattice effect: in fact, without crystal, light reaches the analyser could not be transmitted due to its perpendicular polarization with respect to the analyser polarization axis. Indeed this condition is also named "dark field". On the other hand, when a crystal is introduced, the crystal birefringence produces a light vector rotation of each light wave so that part of the light can pass the analyser.

In the circular polarizer (and in general in the elliptical one), a wave plate is used: it divides the light vector in two orthogonal components at different velocities. Such plate is produced with birefringence materials (Dally & Riley, 1987; Wood, 1964). The wave plate has two principal axes, identified with number 1 and 2 (fig. 24): the transmission of the polarized light along the axis 1 occurs with the velocity c_1, while that one along the axis 2 occurs at c2. In general $c_1 > c_2$, for this reason the axis 1 is the fast axis, while the axis 2 is the slow axis.

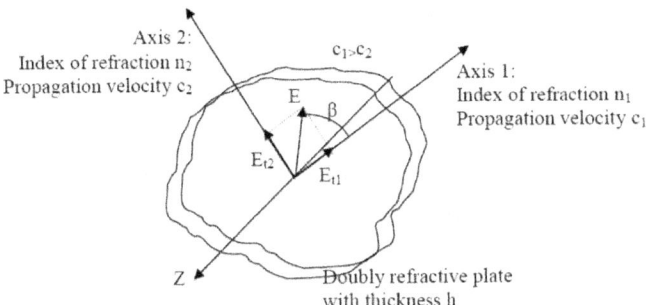

Figure 24: Optical scheme of a wave plate.

If a wave plate is placed after a polarizer, it is necessary to consider that the transmitted wave vector E_t forms an angle β with the fast axis 1. After that it has passed through the plate, E_t is divided in two components E_{t1} and E_{t2}, which are parallel respectively to 1 and 2 axis. The amplitudes of each resulting vector are:

$$E_{t1} = E_t\cos\beta = a\cos\alpha\,\cos\omega t\,\cos\beta = k\cos\omega t\,\cos\beta$$
$$E_{t2} = E_t\sin\beta = a\cos\alpha\,\cos\omega t\,\sin\beta = k\cos\omega t\,\sin\beta \qquad (27)$$

Where $k = a\cos\alpha$. Since the two components travel with different velocities (c_1 and c_2), they cross the plate at different times, implying a relative phase offset. Considering h the plate thickness, the relative delay, that both the wave, travelling across the plate, have with respect to a wave travelling in air (n is considered the air index of refraction), is respectively:

$$\delta_1 = h\,(n_1 - n) \qquad \delta_2 = h\,(n_2 - n)$$

And then the phase difference is

$$\delta = \delta_2 - \delta_1 = h\,(n_2 - n_1)$$

The angular phase difference Δ results:

$$\Delta = \frac{2\pi}{\lambda}\delta = \frac{2\pi h}{\lambda}(n_2 - n_1) \qquad (28)$$

When $\Delta = \pi/2$ the wave plate is called a quarter wave plate ($\lambda/4$). Once the two waves have abandoned the plate, they can be described by the following equations:

$$E_{t1}' = k \cos\beta\cos\omega t$$
$$E_{t2}' = k \sin\beta\cos(\omega t - \Delta)$$

(29)

Recombining the two waves, the amplitude of the resulting wave vector considering these two components is expressed as following:

$$E_t' = \sqrt{(E_{t1}')^2 + (E_{t2}')^2} = k\sqrt{\cos^2\beta\cos^2\omega t + \sin^2\beta\cos^2(\omega t - \Delta)}$$

(30)

The angle with respect to the axis 1 of the plate is:

$$\tan\gamma = \frac{E_{t2}'}{E_{t1}'} = \frac{\cos(\omega t - \Delta)}{\cos\omega t}\tan\beta$$

(31)

In order to obtain a circular polarization the $\lambda/4$ plates are used, with Δ equal to $\pi/4$. In this configuration it is possible to write:

$$E_t' = \frac{\sqrt{2}}{2}k\sqrt{\cos^2\omega t + \sin^2\omega t} = \frac{\sqrt{2}}{2}k$$

$$\gamma = \omega t$$

(32)

It is possible to observe that the amplitude of the light vector is constant, while its direction (which is indicated by the angle γ with axis 1 of the plate) varies linearly with time: therefore, the tip of the vector forms a circle. In particular, if $\beta = \pi/4$ the rotation is counterclockwise, while if $\beta = 3\pi/4$ the rotation is clockwise. In order to obtain an elliptic polarization a $\lambda/4$ plate is used oriented in such way that $\beta \neq n\pi/4$ (with n integer). It is possible to have:

$$E_t' = k\sqrt{\cos^2\beta\cos^2\omega t + \sin^2\beta\sin^2\omega t}$$

$$\tan\gamma = \tan\beta\tan\omega t$$

(33)

therefore, the tip of the light vector forms an ellipse. In general, $\lambda/4$ plate is used in order to obtain the circular polariscope see Fig. 25.

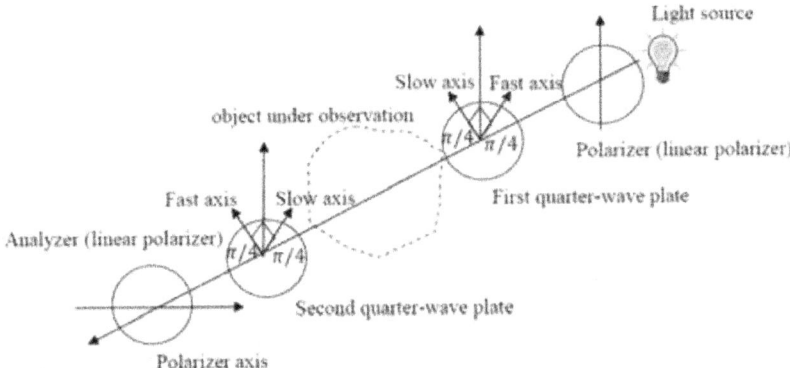

Figure 25: Circular polariscope scheme.

The first element is the polarizer which converts light in linearly polarized light with vertical direction. Then there is the λ/4 plate which is placed with an angle β = π/4 with respect to the polarization axis of the polarizer. In this way light undergoes circular polarization. Another λ/4 plate is place with the fast axis parallel to the slow axis of the previous one: the task is to convert the circular polarization in linear one with vertical direction. As last element, there is the analyzer with horizontal polarization axis which produces the dark field. The presence of the crystal between the two λ/4 plates let light pass through the analyser. In this way it is possible to observe interference fringes. The interference figures belong to two families: isochromatics and isogyres. Intercepting the light coming from the analyser of the polariscope with a screen or plane of the observer, the isochromatic curves represent the loci where all rays with the same difference in optical path strike on such plane, while interference figure where the light vibration directions through the specimen are parallel to the polarization directions of polariscope are the isogyres.

Photoelasticity for Quality Control of Crystal Samples

The use of photoelastic techniques for quality control involves a knowledge of the piezooptical properties of the crystal. As a matter of fact, the number of parameters concerning the piezo-optical effect, as far as refraction index variation cannot be directly calculated without the piezo-optic matrix Π, that relates stress and refraction indices. The components of the piezo-optic matrix Π depend on the symmetry group for each crystal (Nye, 1985, Sirotin et al., 1982). Due to the complexity of the three-dimensional problem of piezo-optic response of scintillating crystals, the values of the single components of Π, at present, are unknown for most crystals (to our best knowledge); therefore the procedure presented hereafter is essentially a semi-emipirical approach, which provides qualitative information and an integral indicator of the internal stress state which we call a quality index. It is not an accurate measurement of internal stress distribution, but nevertheless provides useful information for assessing if residual stress state developed in the crystal has reached critical values.

The methodology for quality control of the internal stress in scintillating crystals has been developed and demonstrated on the uniaxial PbWO$_4$ (PWO) crystal, but it can be extended to the whole class of uniaxial crystals. (PWO) is an optically uniaxial birefringent crystal with ordinary and extraordinary refraction indices n$_o$ = 2.234 and n$_e$ = 2.163 respectively, for λ = 632.8 nm (Baccaro et al., 1997). The development has been carried out on long prismatic samples, cut from an ingot and polished. They can be represented in a (x,y,z) Cartesian coordinate reference system with a solid body having rectangular cross-section (in the x-z plane) and length L (along y axis). The crystallographic c axis coincides with the optical axis (Born & Wolf, 1975; Walhstrom E., 1960), in a stress-free condition, that in the (x,y,z) reference coincides with the z axis. This is also the observation direction. When the crystal sample is subjected to a uniform monoaxial compressive stress σ_y, this compressive stress induces the crystal to became biaxial, and, following the classical interference theory concerning anisotropic crystals, as stated by Born and Wolf (Born & Wolf, 1975), applied to bi-axial crystals, it can be found a fourth-order polynomial expression (34). Eq. 34 (Rinaldi et al., 2009) represents a model for the loci of the interference surfaces called the Bertin surfaces (Walhstrom, 1960):

$$(N\lambda)^2 = \left(x^2 + y^2 + z^2\right) \cdot \left(n_x - n_z\right)^2 \left(1 - \left[\frac{z \cdot \cos\beta + x \cdot \sin\beta}{\sqrt{x^2 + y^2 + z^2}}\right]^2\right) \cdot \left(1 - \left[\frac{z \cdot \cos\beta - x \cdot \sin\beta}{\sqrt{x^2 + y^2 + z^2}}\right]^2\right)$$

(34)

where n$_x$, n$_z$ are the refraction indices along the x and z axes, N is the fringe order, λ is the light wavelength of the light source for the observations, β is the semi-angle between the two optical axes when the crystal becomes biaxial under stress. β is represented by the following function of the three refraction indices n$_x$, n$_y$, n$_z$ (Walhstrom E., 1960). Equation (35) holds for negative crystals (i.e. n$_e$ < n$_o$) like PWO is (Walhstrom, 1960).

$$\tan^2\beta = \frac{\dfrac{1}{n_y^2} - \dfrac{1}{n_x^2}}{\dfrac{1}{n_z^2} - \dfrac{1}{n_y^2}}$$

(35)

For a sample of fixed thickness, equation (34) represents interference images given by "Cassini-like" 4th-order curves in the x-y plane. From equation (34) it can be obtained a family of fringes (isochromatic) that are parameterized by the fringe order N. Attention can be focussed on the first-order fringe (N = 1), as visible in a dark-field configuration of the plane polariscope (i.e. analyser perpendicular to laser polarisation). From a phenomenological point of view, supported by experiment evidence, it can be observed a linear dependence of the refraction index n_y on the applied stress σ, along the direction of application, at least for low stress levels. As a matter of fact, an applied stress along y affects all three refraction indices and the refraction index variations depends on the tensor ΔB (variation of dielectric impermeability) as expressed the matrix equation:

$$\begin{Bmatrix} \Delta B_{xx} \\ \Delta B_{yy} \\ \Delta B_{zz} \\ \Delta B_{xz} \\ \Delta B_{yz} \\ \Delta B_{xy} \end{Bmatrix} = \begin{bmatrix} \pi_{xxxx} & \pi_{xxyy} & \pi_{xxzz} & 0 & 0 & \pi_{xxxy} \\ \pi_{xxyy} & \pi_{xxxx} & \pi_{xxzz} & 0 & 0 & -\pi_{xxxy} \\ \pi_{zzxx} & \pi_{zzxx} & \pi_{zzzz} & 0 & 0 & 0 \\ 0 & 0 & 0 & \pi_{xzxz} & \pi_{xzyz} & 0 \\ 0 & 0 & 0 & -\pi_{xzyz} & \pi_{xzxz} & 0 \\ \pi_{xyxx} & -\pi_{xyxx} & 0 & 0 & 0 & \pi_{xyxy} \end{bmatrix} \begin{Bmatrix} \sigma_{xx} \\ \sigma_{yy} \\ \sigma_{zz} \\ \sigma_{xz} \\ \sigma_{yz} \\ \sigma_{xy} \end{Bmatrix}$$

(36)

where the Voigt notation is used, and the Π components depend on 4/m point group symmetry concerning the PWO (Nye, 1985, Sirotin et al., 1982).

The dielectric tensor [ε] is obtained by the relation:

$$[B] = [B_0] + [\Delta B] = [\varepsilon]^{-1}$$

(37)

In a principal reference system, the refractive indices n can then be derived by:

$$n_i = \sqrt{\varepsilon_i}$$

(38)

Numerical simulations, based only on the variation of tanβ (eq.35) in equation (eq.34), produce results in agreement with the experimental observation in calculating the isochromatic interference fringes. Therefore it appears a possibility to relate internal stress state to fringe geometry; the quality control methods developed are based on this experimental evidence, supported by theory. The interference images in the case of stress free uniaxial samples are families of circles. An applied load on the crystal sample induces a distortion that in the simple uniaxial stress case is a Cassini-like curve (Rinaldi et al., 2009), as the crystal becomes biaxial owing to the applied stress. For low stress level, these curves resembles ellipses.

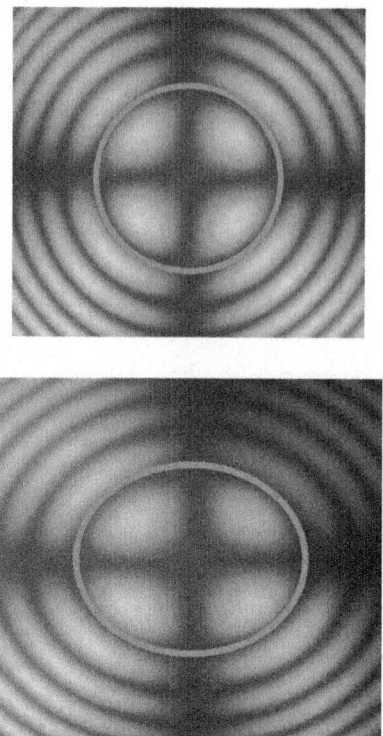

Figure 26: PWO interference images. The highlighted first order fringe is fitted by the model from eq. 34. On the left, the crystal is stress free, on the right, it is subjected to a uniform, uniaxial compressive stress.

Since the evaluation of the refraction index variations by means of (eq. 36) is a hard task, owing to the lack in the knowledge of the Π matrix, an alternative option is to evaluate the fringe distortion by means of an experimental index correlated to fringe distortion; to this purpose it was defined an elliptical ratio Cell (Cocozzella N. et al., 2001) as:

$$C_{ell} = \frac{a}{b} - 1$$

(38)

where a and b are the major and minor axes (along x and y respectively) of the first order isochromatic fringe obtained by observing the crystal in a plane polariscope in dark field (fig. 27). Therefore, in an empirical way, it is established a link between internal stress and fringe distortions by defining the photoelastic constant f_σ (Cocozzella N. et al., 2001) as:

$$\sigma_y \cdot f_\sigma = C_{ell}$$

(39)

In a series of works (Cocozzella N. et al., 2001; Lebau M. et al., 2005), it was experimentally verified in PWO samples that f_σ is a constant for a sample with thickness $z = d$, as Cell depends linearly on σ_y. So it is necessary to systematically evaluate f_σ for PWO samples with different thicknesses d to relate (eq. 39) with the parameter d. Once this photoelastic parameter is known by calibration, which means experimental loading of a crystal sample with known loads, then the same parameter can be used on unloaded samples to assess if an internal stress state is present; the amount of distorsion of the isochromatic fringe provides an empirical assessment on the existence of internal stress.

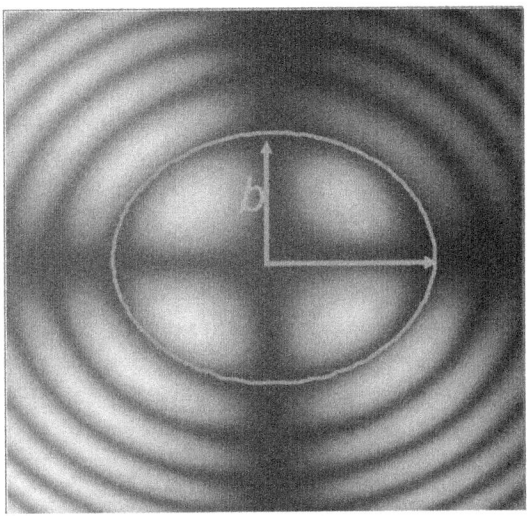

Figure 27: Parameters used to define the elliptical ratio.

In order to know f_σ as a function of d it is necessary to have a set of good quality PWO samples to which a known uniaxial load has to applied. This procedure was described in (Davì & Tiero, 1994); in that case samples have been chosen respecting the "De Saint Venant" conditions with thickness ranges from 5 to 15 mm and a dedicated compression loading machine was used. A dedicated polariscope employing a He-Ne laser source ($\lambda = 632.8$ nm) to perform the quality control tests can be designed (fig. 28) according to the classical polariscope theory (Born M., 1975).

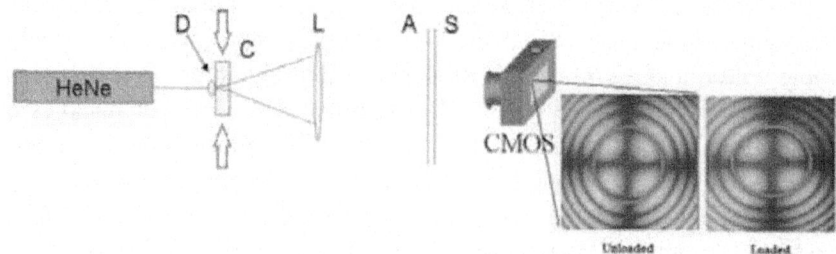

Figure 28: Laser-light plane polariscope. In the dark field configuration, the analyser is set perpendicular to the light polarisation. D = glass diffuser. C = sample. L = convergent lens. A = analyser. S = ground glass screen.

Some changes must be introduced when using laser source instead of a non-coherent diffused light source with respect of classical polariscope (Lebeau et al., 2005; Lebeau et al., 2006; Frocht, 1941). As the laser light is already linearly polarized, the polarizer is not required. Moreover, interference fringes are obtained in convergent light so a small groundglass diffuser was positioned just before the sample. Finally, all the parallel rays emerging from the crystal are focused on a ground glass screen, creating a bijective correspondence between the propagation direction inside the crystal and a point on the screen (Born & Wolf., 1975). C_{ell} is systematically measured versus f_σ applying uniaxial stresses along the y axis crystals, for different thicknesses of the PWO samples. All measured C_{ell} values exhibit a linear trend with the stress σ in the loading range 0-4 MPa. Four cases are reported in figure 29.

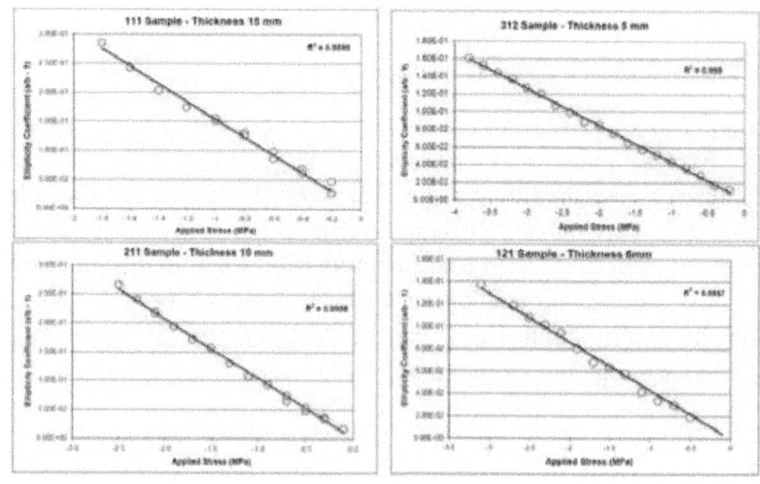

Figure 29: The elliptical ratio Cell as a function of applied stress for the four samples of different thickness.

Numerical simulations, based on equation (34), and experimental data confirm that, Cell linearly scales with the thickness. The resulting f_σ must also follow a linear dependence on crystal thickness z. The experimental data obtained using a laser with wavelength $\lambda = 632.8$ nm, lead to the evaluation of f_σ through a linear regression (correlation coefficient R = 0.997):

$$f_\sigma = 0.0172(\pm 0.0049) - 0.0114(\pm 0.0005) \, z \qquad (40)$$

Numerical simulations from equations (eqs. 34 – 40), using realistic values of n_x, n_y, n_z, confirm the linear behaviour of f_σ versus z at least for z value not too close to z = 0 (for z=0 the analysis loose physical meaning) and for z lower than 20 mm. For larger samples (up to 30 mm), only simulations have been performed , in this case a second-order law rules the variation of f_σ versus z. For z ranging from 5 to 15 mm the linear law can be applied (Ciriaco et al., 2007).

Mapping Residual Stress Distribution in a Crystal Boule

The knowledge of the photoelastic constant f_σ allows the evaluation of the internal stresses for PWO samples by means of the determination of the elliptical ratio, observing the crystals by means of plane polariscope using the same light length. The residual stresses developed during crystal growth tend to increase proportionally to the boule diameter, due to the thermal gradients resulting from growth conditions. The presence of residual stress can hardly be solved by process control. The knowledge of the stress distribution inside the sample during or after growth, can be used as a quality control technique and provide feedback for growth process optimization; furthermore, it can address useful information for planning the mechanical processing.

In order to prove this concept, PWO samples have been studied by the photoelastic method explained above. For each sample, it is therefore determined the ellipticity coefficient C_{ell} of the first isochromatic fringe and through the photoelastic constant it is derived a stress estimate by:

$$\sigma = C_{ell} / f_\sigma \qquad (41)$$

The boule was grown using the Czochralski method, with optical axis orthogonal to the sample axis (fig. 30); 8 samples were cut from the boule, with reference to the X-Y-Z cartesian frame of the figure. The stress distribution has been mapped in the samples. at different locations (x,y,z).

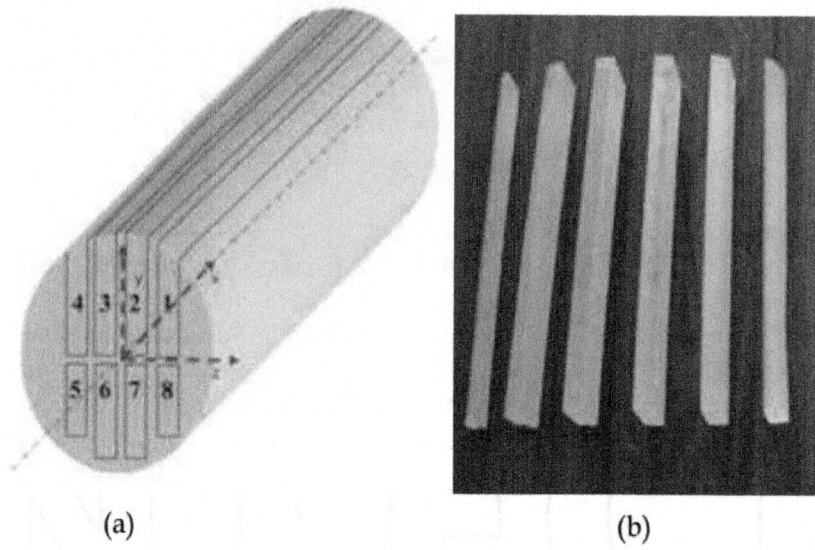

(a)   (b)

Figure 30: Samples position respect to the boule and reference axes (a); a photo of the 8 samples (b).

The mapping gives the possibility of the 3D recostruction of the stress distribution inside the boule; of course, one should taking into account that what observed is a stress state in the samples after cutting, therefore it represents the stress state after relaxation. If an appropriate model would be available, it could be possible to reconstruct the effective stress inside the boule after the growth before the cutting. In fig 31 the data for sample 2 are reported. From the overall data we can deduce that the stress decreases from seed to boule end. As expected, owing to delicate initial growth phase, the classical constant gradient distribution is due to high peripheral tension compared to low internal compression. It should be noted that values are taken after the cutting off of crystal volume in a larger boule; this changes the internal stress state. The original tensional state is unknown and certainly can be higher because of stress relief from cutting. However, the method gives information about the growth of crystals that are fundamental to understanding the growth phenomenon and for the design of the production cycle. Therefore it provides a method for quality control and tuning of crystal production through the analysis of samples taken from the process.

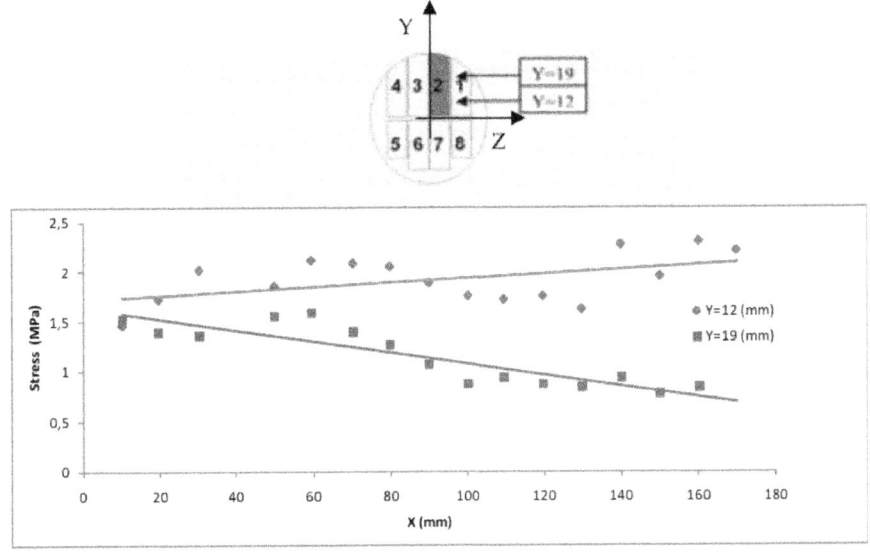

Figure 31: Stress map inside the sample 2 as a function of sample length x, starting from the seed.

A theoretical model of stress is needed in order to extrapolate the initial stress in the boule. Given this limit, this method allows for modifying and optimising the production process. It may lead to re-shaping the boule cone profile to obtain crystals with lower residual stresses. Moreover the obtained data pave the way to balance the process before and after annealing (stress-relief pre-cuts) and for appropriately sequencing the cutting operations, in order to reduce the losses by breaking.

Quality Index

Given the characteristics of the quality control method presented above, it emerges that if a synthetic quality index would be available, the procedure could be helpful in tuning a production process. It is worth to remark again that photoelasticity provides information about stress components in the plane normal to the polariscope axis, aligned in the direction of the optical axis. It is necessary to stress that what is observed, results from the effect of the principal stress difference absolute value, e.g. $|\sigma_{11}-\sigma_{22}|$, in the plane normal to the observation direction. So, we cannot measure, as a matter of fact, the value of each stress component. A single stress component is measured in the case of clearly monoaxial stress field, as in the classical case of an imposed stress by a loading device (Rinaldi et al., 2009, Lebeau et al., 2005) In what follows, for convenience the expression stress-level has been used instead of the absolute principal stress difference. The reported analysis is applied to PWO, but can be extended to the overall family of similar uniaxial crystals. The family of circular isochromatic fringes are in a plane normal to the optical axis, in the case of unstressed crystals. Stress components contained in this plane induce a biaxial state and we detect a family of Cassini-like fringes as explained before, and following the protocol explained in this chapter , the internal stress level was computed from the measurement of the elliptical ratio.

The production of a large number of high quality crystals is a hard goal to perform owing to the complex production route involving different parameters to be controlled. For this, a quality control plan is needed. The need to produce while minimizing costs and production time, leads to the implementation of a fast and easy feed-back on growth parameters, such as temperature distribution and solidification-front velocity. In the following it is proposed a quality feedback for process optimization, obtained by a fast characterization of sample crystals taken from the pre-serial production using the photoelastic methods outlined above. For this purpose, it is possible to use also a classical plane polariscope (in alternative to the previous explained laser light polariscope). In the proposed analysis a green monochromatic light ($\lambda =$ 530 nm) has been used.

Quality control dedicated to the selection of the better production process is made by taking crystals randomly extracted from every delivered batch. In this example the ingots were produced by a "modified" Bridgman method: the platinum crucible has a 35x35 mm2 section and is about 300 mm long. The fusion front temperature is controlled at 1250 (nominal) ± 0.5 °C. A slow and steady shift of the solidification front is produced by the movement of a thermal gradient estimated 30°C/cm at the solid-liquid interface. Using the Bridgman method, the optical axis is along the longitudinal axis, for this, each ingot was cut in 10 mm-thick slices, numbered starting from the seed as shown in fig. 32.

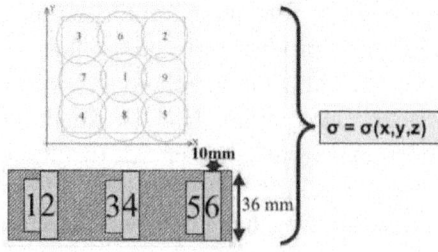

Figure 32: Typical positions observed in each slice. Position 1 is called centre, positions 2-5 corner and position 6-9 "edge middle" (upper figure). Position of the slices cut normal to the crystallographic C-axis (optical axis) (lower figure).

The stress can be measured in typical locations in each slice, as is shown in Fig. 32. The cut of the slices orthogonal to the growth axis (Z), induces a stress relaxation in the axial components σ_z and in the shear stresses τ_{xz} and τ_{yz}.

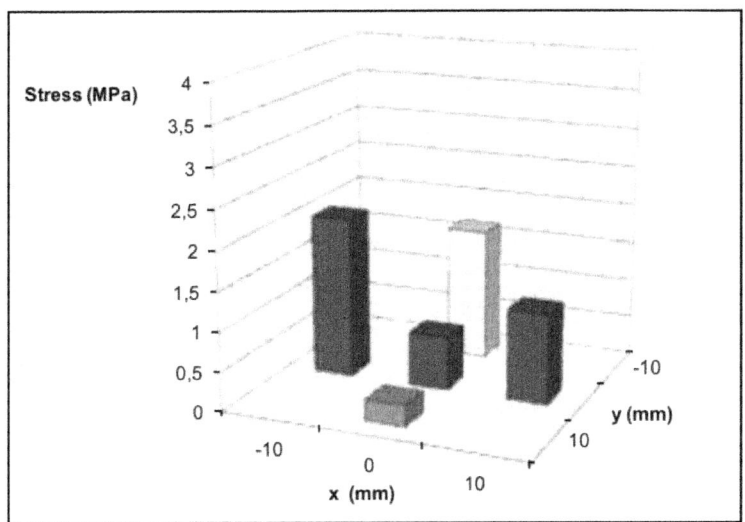

Figure 33: Principal stress along x and y directions on slice 5.

A 3-D reconstruction of stress distribution is possible staring from the measurements for each slice in the typical positions. The reported example leads to the following data (fig. 33, 34 and 35).

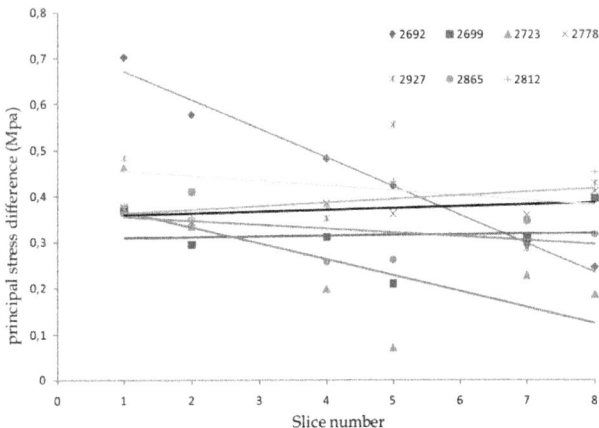

Figure 34: Longitudinal average stress at slice corners for the 7 samples labeled in the figure. Generally decreasing trend from the seed to the end of crystal is confirmed. Linear regression have been shown for each crystal.

The large stress difference at the corners for Sample 2692 in the first slice can suggest the presence of thermal problems after the solidification in the initial part or a crucible defect. The single crystal stress homogeneity seems to be restored at the crystal end. A not optimal thermal field displacement can also be supposed. The speed of temperature change could have been too high in the early growth stages of 2692 sample.

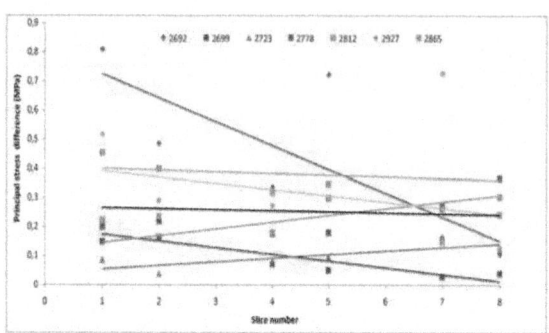

Figure 35: Longitudinal trend at centre of the slices for the 7 samples labeled in the figure.

The data collected can provide an analytical overview of the problem by assessing the evolution of stress in the mapping of the crystals. This analysis puts emphasis on individual problems that emerged during growth. The need of a synthetic index of quality derives from the necessity to understand what production parameters must be changed in the direction of better overall quality. The problem lies in handling a large quantity of data, as in the example shown. Extracting the average measures of stress from 6 samples we can write the following table:

Table 1: Mean stress and standard deviation of 6 samples

SAMPLE	St. Dev. [MPa]	Mean [MPa]
2692	0.38	0.62
2699	0.38	0.36
2723	0.37	0.29
2778	0.33	0.54
2812	0.23	0.52
2865	0.33	0.39

Quality control dedicated to the selection of the better production process is made by taking crystals randomly selected from every delivered batch. A large set of data can be collected from ingots produced in different conditions.

The residual stress inside the crystals can vary heavily depending on the type of process used but also as a function of the position inside the crystal. In order to identify the best method of production and to guide the development of production methods, it's necessary to find a summary measure of the quality of the samples. To evaluate in synthetic manner the overall quality of each crystal, the data can be put in a Cartesian plane, with X axis the average stress level (σ_{av}) of all the measured points from each ingot, and Y axis the corresponding standard deviation (S). It is possible to define an "index of quality", R (Rinaldi et al., 2010):

$$R = \sqrt{(kS)^2 + \sigma_{av}^2}$$

(42)

In brittle materials, as in particular single crystals, the stress level variation may cause breaking risk, due to the gradient, as dangerous as the than an high average value σ_{av}. For this, larger values of R correspond to the lower crystal quality. The standard deviation S is weighted with a coefficient k (≥ 1) to be set from the producer in order to achieve the results identified on the basis of the required quality standards. k amplifies the standard deviation with respect to the average stress value highlighting the existence of gradients and inhomogeneity.

The general purpose of the (eq. 42) need some explanations linking the mathematical properties of (eq. 36) to the physical meaning of the parameters. In the plane S-σ_{av}, the R=const. is a boundary delimiting an area related to the ingots quality. Each crystal is identified by its σ_{av} and S. The accepted crystals are in the area below the curve R=const. From the producer point of view, R will be the maximum σ_{av} acceptable (that is σ_M). Choosing k following the experimental experience and related to the quality, in terms of homogeneity required, it is easy to calculate the maximum S, that is $S_M = R/k = \sigma_M/k$. It is evident that the maximum standard deviation, S_M, must be lower than, or at least equal to the average stress level in the limit case. From the quality point of view the need that the stress variation must be controlled leads to a new constrain:

$$S = c\sigma_{av}$$

(43)

with c the slope of a line in the plane S-σ_{av}, c must be chosen for c ≤ 1, as explained above. The acceptance area is below the curves. These constrains depend both from the average stress and the weighted stress standard deviation. The condition that the standard deviation cannot exceed the average stress value is thought fundamental in brittle materials, in fact, the stress variation, due to the crystal heterogeneity, might increase the breaking risk. In fig. 36 an example of this technique is shown.

The boundary curves delimiting the acceptability area are obtained with k = 1.5 and the limit quality value is R = 0,7 MPa. c = 1 was chosen as a limit case. These parameters can be set by the manufacturer according to statistical data of accidental cracking. Only two samples (highlighted by the blue oval) lie in the acceptability area, that is below the curves 42 and 43. The fact that three samples (2865, 2723 and 2699) are close to the line where standard deviation is equal to the average stress value clearly highlights the existence of high stress gradient. In particular the 2723 and 2699 samples are below the curve (eq. 42) but above the line (eq. 43), therefore are not accepted due to the high stress gradient. From this analysis it is possible to conclude that the process 2865 and 2812 have the best production parameter and indicates the development direction to improve the crystals quality. As final analysis it is possible to perform a comparison between the best and worst samples to put in evidence the specific critical points, as shown in the figure 37 (Rinaldi et al., 2010).

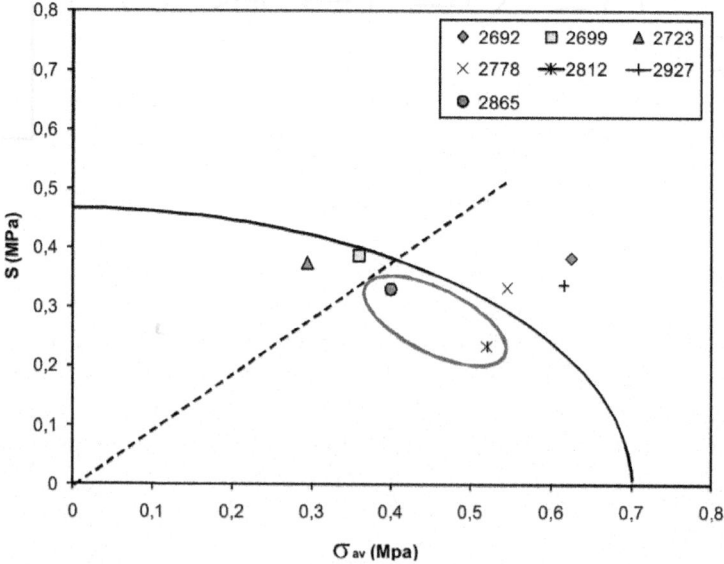

Figure 36: Different samples in the plane S-σ_{av}. Only samples 2865 and 2812 are accepted, meeting the quality requirements.

Fig. 37 clearly put in evidence that the low quality sample 2692 exhibits high absolute stress values but also high stress variation. On the other hand, the higher level quality sample 2865 has lower absolute stress values and it appears more homogenous. It indicates that the specific production process is well tuned and that probably the production parameter and gradients are well controlled yielding an homogeneous sample.

CONCLUSIONS

Scintillating crystals are widely used in radiographic systems, in computerized axial tomography devices and in calorimeter used in high energy physics. Scintillating crystals are cut to their final shape from an ingot, which is grown by classical crystal growth techniques. From a mechanical point of view, the quality of a crystal is closely related to its geometry, to the surface finish and moreover to its internal state of residual stresses. In particular an excessive residual stress is a major cause of crystal breakage, which often may occur during crystal cut, during surface finishing or, even worse, only when the crystal is assembled into the detector units.

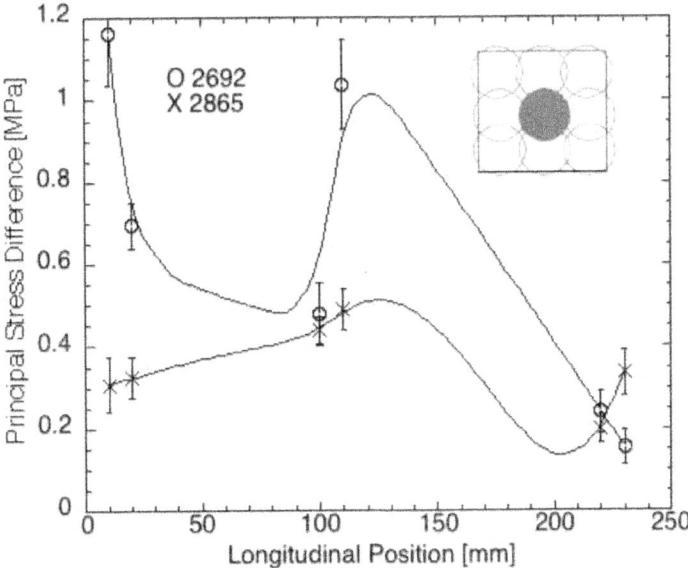

Figure 37: Comparison between better/worst samples at centre position of the slices (as shown by the inset) as a function of the longitudinal position from the seed.

The need to produce high-quality crystals is therefore fundamental both to avoid damage during assembly and finishing of crystals. Crystal performance in terms of production of light strongly depends on surface finish, therefore crystal tool machining is a crucial process to achieve the high performance needed in the case of scintillating crystals for high energy physics and medical applications.

For optimal crystals performance, attention has therefore to be paid to the mechanical aspects of the production process; from the mechanical point of view this can be guaranteed by adequate quality control methods. If adequate quality inspection of crystals is achieved, this has the potential to prevent breaking during the assembly in an array. The authors have reported the experience which was made within the collaboration with CERN to the development of the electromagnetic calorimeter of the Compact Muon Solenoid (CMS) presently working at CERN. From an industrial point of view, the trend is to use smaller and smaller crystals for biomedical instrumentation; in such crystals the surface plays an even more relevant role in the production of light. For this reason, the final mechanical processing is important for producing high quality crystals. Therefore the experience made for the large crystals of CMS is in general valuable to guide the development of suitable quality control methods for scintillating crystals and in particular for biomedical industry.

An increasing attention to limit production costs requires an assessment of crystal quality by a fast and possibly non-destructive methodology, finalized to tune and keep under control the crystal growth and finishing processes, and to eliminate from the production process the crystals which are produced out of tolerance, thus reducing downtime and waste. Internal residual stress is not only the most important causes of breaking, but may be interpreted as an overall quality indicator.

Residual stresses, induced by temperature spatial and temporal distribution during the growth and by complex interaction of the melt material and the growing ingot with the crucible, play an important role in production yield in terms of cracking risk during mechanical processing and heterogeneity in finished crystal properties. A regular production of good crystals requires a quality control plan leading to a fast and easy feedback on growth parameters, such as temperature distribution and solidification-front velocity. The developed methodology for quality control consists in providing the producer a quality feedback for process control and optimization, obtained by experimental characterization of sample crystals taken from the pre-serial production by photoelasticity. Photoelasticity is a measurement technique aiming to study and evaluate the stress state inside a transparent medium. In traditional photoelasticity a plane stress state distribution is studied, by means of a plane polariscope. Usually it is applied to optically isotropic media, Perspex, glass or optically isotropic crystals, which become birefringent under stress.

Referring to naturally anisotropic media, such as uniaxial and biaxial scintillating crystals, the observation of unstressed crystals, by means of a plane polariscope, shows a symmetrical interference pattern due to the symmetry of the lattice. An internal stress state induces a lattice symmetry distortion.

The modelling of the interference image obtained from an anisotropic uniaxial crystal when a stress state is present, and the measurement of characteristic parameters of the interference fringe pattern offers a mean for quality control able to provide spatially integrated information on the internal stress.

Although a mathematical modelling of the piezo-optical effects is possible, the knowledge of the coefficients of the model is not complete and accurate; therefore a semi-empirical approach is proposed. This leads to the definition of a parameter correlated to the deformation of the fringe pattern of a crystal under stress. The ellipticity, introduced into the fringe pattern is due to the stress state. Linear regression of experimental data of ellipticity vs. stress, collected with crystals undergoing known stress states, allows to build an experimental relationship which can then be used for quality assessment of unknown crystal samples. If the internal stresses are residual stresses, this allows to develop a quality control method to detect the presence of residual stresses non invasively. The method could be applicable on samples taken from the production, for process optimization and control, or it can be applied on the finished crystal as a pass-fail filter for removing from the batch all samples which exceed prescribed limits.

The statistical analysis of many data from samples randomly taken from a pre-serial production allows to build a quality index depending on mean stress value and on its standard deviation, which are quantities related to residual stress intensity and gradient. This index can be used as a global indicator of process capacity to produce crystals with acceptable residual stress state.

This method suggests therefore a quality indicator to synthetically evaluate the production by means of a criterion of acceptability, useful in general crystal production. The procedure and the quality index have been validated on PbWO4 (PWO) uniaxial scintillating crystals; they have been intensively studied owing to the necessity of large amount of them (about 82000 large crystals) for the CMS. In fact the production effort needed a fast and reliable quality control.

In that case study, the attention was focused on the measurement of residual stresses over the whole crystal volume, particularly in sections cut perpendicularly to the optical axis. The collected data enabled the construction of a 3-dimensional stress map for each crystal from a pre-serial production. The detection of internal stress and defects, can be related to the corresponding production parameters and may suggest improvements in the production process or highlight criticalities to be solved before a serial production is started. What presented is demonstrated for uniaxial crystals, but the same approach can be extended to all types of crystals, particularly those of a new generation (LYSO, LuYAP) as a function of their applications in high energy physics and for medical diagnostics. As conclusive remarks, we have to consider that other techniques should be taken into account to analyse crystals quality. In particular researchers are paying attention to experimental methods for the assessment of the surface damage, which are not treated in this chapter: X-ray diffraction (XRD), grazing incidence X-ray diffraction (GID) and RX reflectometry (XRR), (Mengucci et al., 2005).

ACKNOWLEDGMENT

This work has seen the contribution of many colleagues, amongst which we thank prof. Giuseppe Majni and Prof. Fabrizio Davì, who contributed through many fruitful discussions. A relevant part of the work has been developed with the direct contribution of students amongst which we warmly thank Nicola Cocozzella, who was the first to deal with this topic, and PhD candidates, in particular dr. Andrea Ciriaco, whose PhD thesis constitutes a milestone in our work.

REFERENCES

1.   Auffray E., Cavallari F., Lebeau M., Lecoq P., Schneegans M., Sempere-Roldan P. (2002). Crystal conditioning for high-energy physics detectors, Nuclear Instruments and Methods in Physics Research Section A (NIM A) 486, pp. 22-34.

2.   Baccaro S., Barone L. M., Borgia B., Castelli F., Cavallari F., Dafinei I., de Notaristefani F., Diemoz M., Festinesi A., Leonardi E., Longo E., Montecchi M., Organtini G. (1997). Ordinary and extraordinary complex refractive index of the lead tungstate (PbWO4) crystal. Nuclear Instruments and Methods in Physics Research Section A, 385, pp. 209-214.

3.   Born M., Wolf E., (1975). Principles Of Optics, 6th ed., Pergamon press, New York, USA. Cocozzella N., Lebeau M., Majni G., Paone N., Rinaldi D. (2001). Quality inspection of anisotropic scintillating lead tungstate (PbWO4) crystals through measurement of interferometric fringe pattern parameters. Nuclear Instruments and Methods in Physics Research Section A (NIM A) 469 3 pp.331-339.

4.   Ciriaco A., Davì F., Lebeau M., Majni G., Paone N., Pietroni P., Rinaldi D. (2007). PWO photoelastic parameter calibration by laser-based polariscope. Nuclear Instruments and Methods in Physics Research A 570, 55–60

5.   Dally J. W., Riley W. F., (1987). Experimental Stress Analysis, 2nd ed., McGraw-Hill Book Company, Singapore.

6. Davì F. and Tiero A., (1994). The Saint-Venant's problem with Voigt's hypotheses for anisotropic solids. J. Elasticity 36, pp. 183-199.

7. Frocht, M.M. Photoelasticity, Wiley, New York, 1941.

8. Hodgkinson I. J., Wu Q. H., (1997). Birefringent Thin Films and Polarizing Elements, World Scientific, New Jersey, USA.

9. Hofstadter, R (1949). The detection of gamma-rays with thallium-activated sodium iodide crystals. Phys.Rev. 75, pp. 796-810.

10. Ishii M., Kobayashi M. (1996). Mechanical properties of PWO. Nuclear Instruments and Methods in Physics Research Section A (NIM A) 376, pp. 203-207

11. Lebeau M. (1985). Monocrystalline bismuth germanate Bi4Ge3O12 (BGO) recent results on mechanical properties. J.Mat.Sci.letters 4, 779-782.

12. Lebeau M., Ciriaco A., Gobbi L., Majni G., Paone N., Pietroni P., Rinaldi D. Quality monitoring in PWO scintillating crystal production during R&D phase Proceedings of the 8th International Conference on Inorganic Scintillators and their Use in Scientific and Industrial Applications, Publisher: National Academy of Sciences of Ukraine, Kharkov (2006) 334-337.

13. Lebeau M. (2003). Crystal Growth Technology. In Methods and Tools for Mechanical Processing of Anisotropic Scintillating Crystals, pp.561-586. Wiley and Sons, London.

14. Lebeau M., Pietroni P., Gobbi L., Majni G., Paone N., Rinaldi D. (2005). Mapping residual stresses in PbWO4 crystals using photoelastic analysis., Proceedings of Scint'03 7th International Conference on Inorganic Scintillators, September 8-12, 2003, Valencia, Spain. NIM A537 154-158

15. Lecoq P. et al. (2006). Inorganic Scintillators for Detector Systems. ISBN-10 3-540-27766-8 Springer Berlin Heidelberg New York.

16. Mengucci P., Di Cristoforo A., Lebeau M., Majni G., Paone N., Pietroni P., Rinaldi D. (2005) Surface quality inspection of PbWO4 crystals by grazing incidence X-ray diffraction. Nuclear Instruments and Methods in Physics Research Section A (NIM A) 537, 207- 210.

17. Perelomova N. V. and Tagieva N. M., (1983). Problems in Crystal Physics with solutions, Mir Publishers, Moscow, Russia.

18. Pietroni P., Lebeau M., Majni G., Paone N., Rinaldi D. (2005). Development of Young's modulus non-destructive measurement techniques in non-oriented CeF3 crystals. Nuclear Instruments and Methods in Physics Research Section A (NIM A) 537, 203-206.

19. Rinaldi D., Lebeau M., Majni G., Paone N. (1997). Photoelasticity for the investigation of internal stress in BGO scintillating crystals. Nuclear Instruments and Methods in Physics Research Section A (NIM A) 317-322.

20. Rinaldi D., P. Pietroni, F. Davì (2009). Isochromate fringes simulation by Cassini-like curves for photoelastic analysis of birefringent crystals. Nuclear Inst. and Methods in Physics Research, A 603, 294–300

21. Rinaldi D., Ciriaco A., Lebeau M., Paone N. (2010). Quality control on pre-serial Bridgman production of PbWO4 scintillating crystals by means of photoelasticity Nuclear Inst. and Methods in Physics Research, A 615, 254–258

22. Walhstrom E.E., (1960). Optical Crystallography, Wiley, New York, (USA).

23. Weber M., Monchamp R. (1973). Luminescence of Bi4Ge3O12- Journal of Applied Physics 44: 5495-5499.

24. Wood E. A. (1964). Crystal And Light, Van Nostrand Company, New Jersey.

25. Wooster, W. A. (1938). A test-book on Crystal Physics, Cambridge University Press.

Chapter 8

RADIO FREQUENCY QUADRUPOLE ACCELERATOR: A HIGH ENERGY AND HIGH CURRENT IMPLANTER

Yuancun Nie[1,2], Yuanrong Lu[2], Xueqing Yan[2] and Jiaer Chen[2]

[1]State Nuclear Power Technology R&D Center, Beijing

[2]State Key Lab of Nuclear Physics and Technology (Peking University), Beijing China

INTRODUCTION

In this chapter, issues of applying the radio frequency quadrupole (RFQ) accelerator to ion implantation for material modification and microelectronics will be discussed at first. Subsequently, several examples of the RFQ-based implanters will be introduced, with an emphasis on the ladder IH-RFQ accelerator at Peking University, of which the output beam energy spread is optimized to be much lower than the conventional value by means of a novel beam dynamics design strategy. The chapter is divided into 6 sections, which are the introduction, advantages of RFQ as implanter, problems and solutions of RFQ-based implanters, several RFQs aiming at ion implantation, ladder IH-RFQ at Peking University, and summary.

RFQ accelerator, proposed by I. M. Kapchinsky and V. A. Teplyakov in 1970, can simultaneously focus, bunch and accelerate low energy beam extracted from ion source directly, based on the rf electrical field of a modulated quadrupole transport channel[1]. The details of its principle and development could be found in many literatures[2,3]. Taking advantages of its high current, compact size and good ion selectivity, etc, RFQ has been used widely in the field of particle accelerator, after 40 years' development both in the beam dynamics design and in the rf cavity structure [4]. At present, RFQs can provide high current (up to hundreds mA) ion beams at energy level of MeV over the mass range from Hydrogen to Uranium and the 5 MHz to 500 MHz frequency range[5] .

ADVANTAGES OF RFQ AS IMPLANTER

Many efforts have been made to apply the RFQ accelerator to ion implantation. Generally speaking, the beam line of the RFQ-based implanter consists of the ion source, low energy beam transport (LEBT), RFQ accelerator and the target, equipped with power system, controlling system, vacuum and cooling systems, etc. The advantages of RFQ-based implanter are concluded as follows:

- The beam extracted from the ion source is accelerated further by the following RFQ, so the beam energy can reach the order of several MeV, which fulfills the requirement of deep deposition.

- The current of heavy ion beam accelerated by the RFQ could be in the mA-range. For an implant dose of 10^{18} ions/cm^2, the implantation time needed is less than 3 minutes/cm^2 using an ion beam with average current density of 1 mA/cm^2.

- RFQ can simultaneously accelerate positive and negative ions with the same charge to mass ratio[6]. It makes full use of the rf power, and more importantly it avoids the charge accumulation while doubling the implant flux density.

- For a given RFQ, it is available not only for the designing ion beam, but also for other ion species, as long as the injection energy and the inter-electrode voltage are adjusted appropriately. For example, the 36 MHz IH-RFQ in the UNILAC accelerator of GSI can accelerate all the heavy ions with charge to mass ratio greater than 1/65 from 2.2 keV/u to 120 keV/u[7] .

- Compared to the electrostatic accelerator, high voltage breakdown and high energy Xray radiations are reduced greatly for this case, because high voltage terminal is avoided and the rf inter-electrode voltage in RFQ is normally below 100 kV.

- The equipment is very compact. For instance, the 26 MHz ISR-1000 RFQ at Peking University can accelerate $^{16}O^+$ beam from 22 keV to 1 MeV within 2.6 m, and the cavity diameter is 0.75 m[8] .

PROBLEMS AND SOLUTIONS OF RFQ-BASED IMPLANTERS

The RFQ-based implanter has some drawbacks, i.e. the output beam energy is normally fixed for a given RFQ, and the output beam energy spread is quite big (usually the full width at half magnitude, FWHM energy spread is around 3%).

To make the output beam energy variable, two methods have been used, and they are:

- Making the resonant frequency of the RFQ cavity variable. It is mostly applied to the 4- rod RFQs. For instance, at Frankfurt University several VE (Variable energy) RFQs have been developed, and the output energy range could be doubled and even more while the ground plate connected to the stems is moved to vary the frequency (VF)[5]. Hitachi Ltd. also invented a variable energy RFQ linac with an external tunable one-turn coil, and the variable resonant frequency range is from 9.6 to 39 MHz, which corresponds to an energy variation of 16 times[9] .

- Changing the output energy by using a high voltage platform (HVP) or a Spiral Loaded Cavity (SLC). The variation in the beam energy is continuous in the range determined by the HVP, for example in Ref. [10], the 60 MHz RFQ implanter with the HVP at voltage higher than 250 kV could provide a beam energy variable from 500 keV to 1.5 MeV for ions with charge to mass ratio 2/16. At Frankfurt University, a SLC was coupled to the RFQ designed for the acceleration of N^+ from 50 keV to 1.5 MeV to vary the beam energy in a range of ±200 keV[11] .

To reduce the output energy spread, non-adiabatic bunching method could be used and there are also two ways:

- The first one is the external bunching method. A beam dynamics design with energy spread 0.6% has been realized for $^{14}C^+$ beam using this method[12]. However, the prebuncher used to produce a pulsed beam from a DC beam before the RFQ makes the system complex, and the total beam transmission efficiency is constrained by it.

- The second one is the internal discrete bunching method. Most recently, the low energy spread beam dynamics design for ions with charge to mass ratio greater than 1/14 wasaccomplished[13]. Optimization procedure of the design as well as its preliminary beam experiment will be discussed in detail.

SEVERAL RFQS AIMING AT ION IMPLANTATION

In the following Table 1, parameters of several typical RFQs for ion implantation in the world are summarized.

Table 1: Parameters of some RFQs for ion implantation (after Ref.[14])

Affiliation	RFQ structure	Ions or charge-to-mass ratio	Energy (MeV/u)	Current (mA)	Frequency (MHz)	Length (m)	Note
GSI	Split coaxial	Ar+,Kr+	0.045	5	13.5	9.4	Operation
Hitachi	Inductance-Capacitance	P+,N+	0.043	5	17.3	2.3	Operation
Hitachi[9]	4-rod	1/31	0.036	1	21	2.3	VE(VF)
Peking University	Integral Split Ring	O+,O-, N+	0.0625	0.5	26	2.6	Operation
Nissin	4-rod	B+,N+,O+	0.0835	0.2	33.3	2.22	Operation
Seoul[10]	Modified 4-vane	2/16	0.0625±0.0313	2	60	1.57	VE(HVP)
Simazu	4-vane	B+,N+,P+	0.09	0.1-0.4	70	2.1	Operation
Frankfurt	4-rod	B+	0.09	15	81	1.2	Operation
Frankfurt[11]	4-rod	1/14	0.107±0.014	10	108.5	2.0	VE(SLC)

LADDER IH-RFQ AT PEKING UNIVERSITY

Since the year 2008, a 104 MHz ladder type IH-RFQ has been developed at Peking University, for the acceleration of the ion species with mass-to-charge ratio smaller than 14 from 2.86 keV/u to 35.71 keV/u. As a specific feature, the output beam energy spread is as low as 0.6% achieved through the internal discrete bunching method, which makes potential applications of the RFQ feasible, such as accelerator mass spectrometry (AMS) and ion implantation. Design procedures and the most recent experimental results of the ladder IHRFQ will be described in detail.

Beam Dynamics Design of the Ladder IH-RFQ

In recent years, RFQ based ^{14}C AMS application has been studied extensively at the Institute of Heavy Ion Physics (IHIP), Peking University. Great efforts have been made to reduce the output FWHM energy spread of the RFQ. Non-adiabatic bunching method could be used owing to the lack of space charge effect since the current is usually weak. A beam dynamics design with energy spread 0.6% has been realized by means of the external bunching method. However, the total beam transmission efficiency is constrained by the pre-buncher used to produce a pulsed beam before it is injected into the RFQ. For this reason, we tried another possibility, the internal discrete bunching method proposed by J. W. Staples at LBNL to reduce the output energy spread without the pre-buncher[15]. The final low energy spread beam dynamics design for $^{14}C^+$ RFQ was accomplished. Optimization procedure of the design along with its performances will be reported.

The operating frequency 104 MHz was chosen because of the limitation of the existing rf power transmitter, and our interest of investigating the ladder type IH-RFQ structure [13], in which the four electrodes are supported alternately by erect stems mounted to the external cavity from top to bottom equidistantly, different from the traditional 4-rod RFQ and IHRFQ structures. Properties of the ladder IH-RFQ will be studied in the next section.

According to the strategy of the internal discrete bunching, the whole RFQ beam dynamics design is divided into five sections, different from the conventional four-step method developed at LANL [16], consisting of the radial matcher, the buncher section, the drift section, the transition section and the acceleration section. With the buncher and drift sections serving as an internal non-adiabatic buncher, the new method reduces the output longitudinal emittance significantly and transverse emittance slightly compared with the traditional one. However, neither this strategy nor the conventional method takes into account the match between the beam size and the acceleration channel carefully, leading to beam envelope mismatch and even particle loss. According to the beam matching equation for weak current

$$\varepsilon_{tn}^2 = \frac{a_0^4 \gamma^2 \sigma_t^2}{\lambda^2}$$

(1)

where ε_{tn} is the normalized rms (root mean square) emittance, a_0 is the transverse rms beam radius, g is the relativistic gamma, l is the rf wavelength, and σ_t is the transverse phase advance without current. σ_t can be obtained by solving the transverse motion equation, i.e. the Mathieu Equation, and we have

$$\sigma_t^2 \propto \frac{B^2}{8\pi^2} + \Delta$$

(2)

where B is the focusing parameter for the RFQ accelerator, while D is the defocusing factor, which are defined as follows

$$\Delta = \frac{qe\pi^2 AV \sin\varphi_s}{2m_0 c^2 \beta_v^2 \gamma}$$

$$B = \frac{qe\lambda^2 FV}{m_0 c^2 a^2 \gamma}$$

(3)

where A, V, φ_s, F are the acceleration coefficient, inter-electrodes voltage, synchronous phase and focusing factor, respectively, β_v is the relativistic velocity, a is the minimum aperture radius between two opposite electrodes of one acceleration cell. Based on the matching design method [17], we enhance the focusing parameter B gradually as the defocusing factor $|\Delta|$ increasing due to the growing acceleration coefficient A, during the first stage of the transition section that serves as the shaper section in the four-step method. Nevertheless, in the acceleration section, B is decreased slowly since $|\Delta|$ is reduced because of the rapidly increasing beam energy. The above improvements make the transverse phase advance σ_t constant, thereby the beam matched, accordingly suppress the emittance growth and minimize the particle loss.

In addition, there is an approach to reduce the RFQ energy spread furthermore, which can be shown by

$$\Delta w_{max} = \left[2\xi eAVW_s \left(\varphi_s \cos\varphi_s - \sin\varphi_s \right) \right]^{1/2}$$

(4)

where Δw_{max} is the separatrix height that indicates the maximum beam energy spread, x is the charge-to-mass ratio q/A_0 of the ion, e is the charge of an electron, and W_s is the kinetic energy of the synchronous particles. It can be seen that the energy spread can be reduced by decreasing A and the inter-electrode voltage V while increasing the synchronous phase φ_s, where lower A means lower electrode modulation m. It means that φs should be as large as possible even close to 0° at the last acceleration cell in order to achieve low energy spread, not necessarily be -30° as usual. For our case, the frequency 104 MHz is relatively high for ^{14}C ions with $q/A_0=1/14$, which destines a small aperture radius a in order to obtain sufficient focusing strength under a certain V. Small aperture radius means small electrodetip radius r_0 to avoid excessive Kilpatrick coefficient[2]. To relax the manufacture pressure of the electrodes, we have to enhance V to increase a, unfortunately, it conflicts with reducing not only the energy spread but also the power consumption. As a result of the compromise, an inter-electrode voltage 60 kV is used.

The main parameters were generated by the matched and equipartitioned design code MATCHDESIGN developed at IHIP[17], while the non-adiabatic buncher section was designed by the code of Staples[15]. Beam transport simulation was performed by the program RFQDYN derived from PARMTEQ[16]. In order to make a as large as possible, we tried to reduce B during the transition section and the acceleration section while ensuring the balance between defocusing and focusing of the beam. It finally induced 2% particle loss at the transverse direction. The final value of the synchronous phase φ_s was set to be -6° at the end of the acceleration section, which decreased the output beam energy spread and additionally quickened the acceleration process. The modulation parameter m was also evolved carefully, in view of that large modulation is advantageous to particle acceleration, but disadvantageous to reducing the energy spread.

Figure 1 shows us the evolutions of the focusing parameter B as well as the acceleration coefficient A along with the cell number for the final beam dynamics design. The other dynamics parameters of the RFQ are plotted in Fig. 2. The rapidly advance of φ_s leads to a small portion of particles lose their balances of the longitudinal motion and can't be accelerated effectively. This explains why a tail appears in either the output energy spectrum or the output phase spectrum. The resultant transmission is 97.6%, while 80% particles are in the phase range of $\pm 15°$ around φ_s, and 90% for $\pm 30°$. Meanwhile, more than 90% particles are limited by the energy range ± 15keV, with the FWHM energy spread 0.6%. The total length of the electrode is about 1.1 m to accelerate the $^{14}C^+$ beam from 40 keV to 500 keV, which is supposed to be slightly longer than that of the external bunching method due to the implanted non-adiabatic buncher. In view of the overall simplified system and much higher transmission efficiency in total, the internal non-adiabatic bunching strategy is a competent candidate to design the RFQ for AMS application.

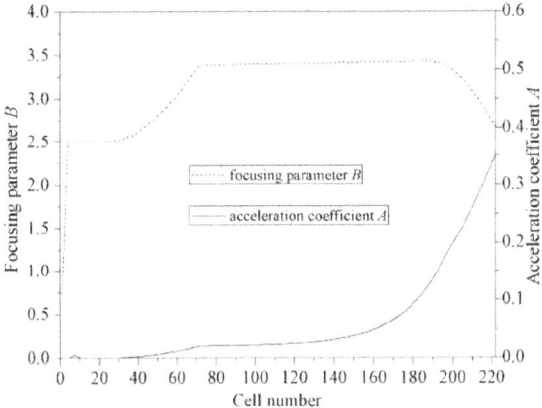

Figure 1: Variations of the focusing parameter and the acceleration coefficient with the cell number of the ladder IH-RFQ.

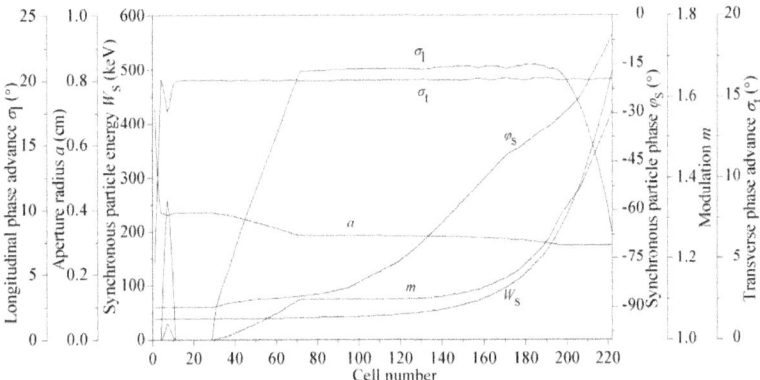

Figure 2: The main beam dynamics parameters varying with the cell number of the ladder IHRFQ.

Responses of the beam transmission efficiency and the output beam energy spread to input Twiss parameters (α, β), current and emittance (normalized, rms) are presented in Fig. 3, with the design values (0.31, 2.19 cm/rad), 0 mA, 0.042 mm-mrad, respectively. Figures 3(a) and 3(b) illustrate that neither the beam transmission efficiency nor the output beam energy spread is sensitive to the Twiss parameters, since all the fluctuations are acceptable within the plotted ranges. The influences of the input emittance on the transmission efficiency and output energy spread are observed but tolerable considering that even when it is twice the design value, the transmission efficiency and energy spread are still higher than 90% and below 1.5%. A remarkable feature is that the transmission efficiency and energy spread are better than 93% and 0.8% at the current 1 mA, despite the beam dynamics was optimized at 0 current.

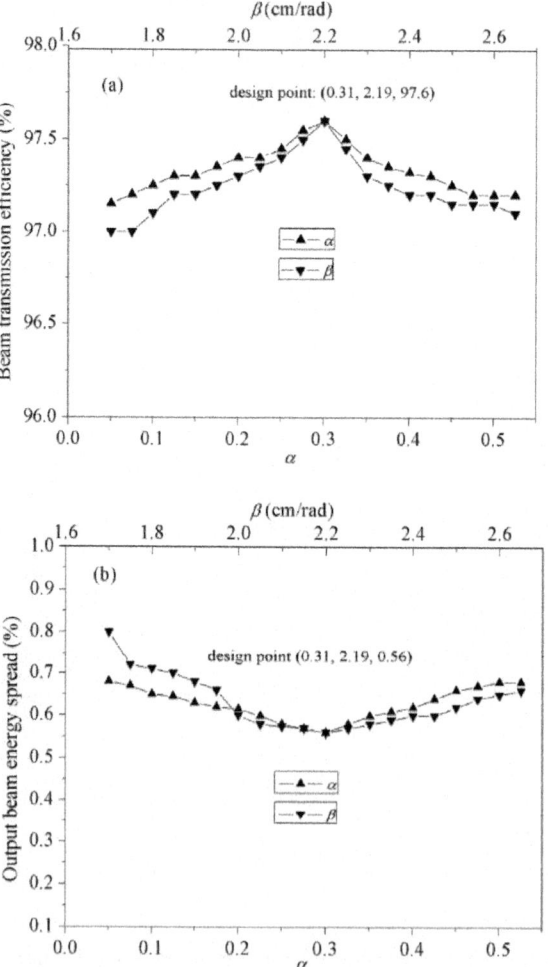

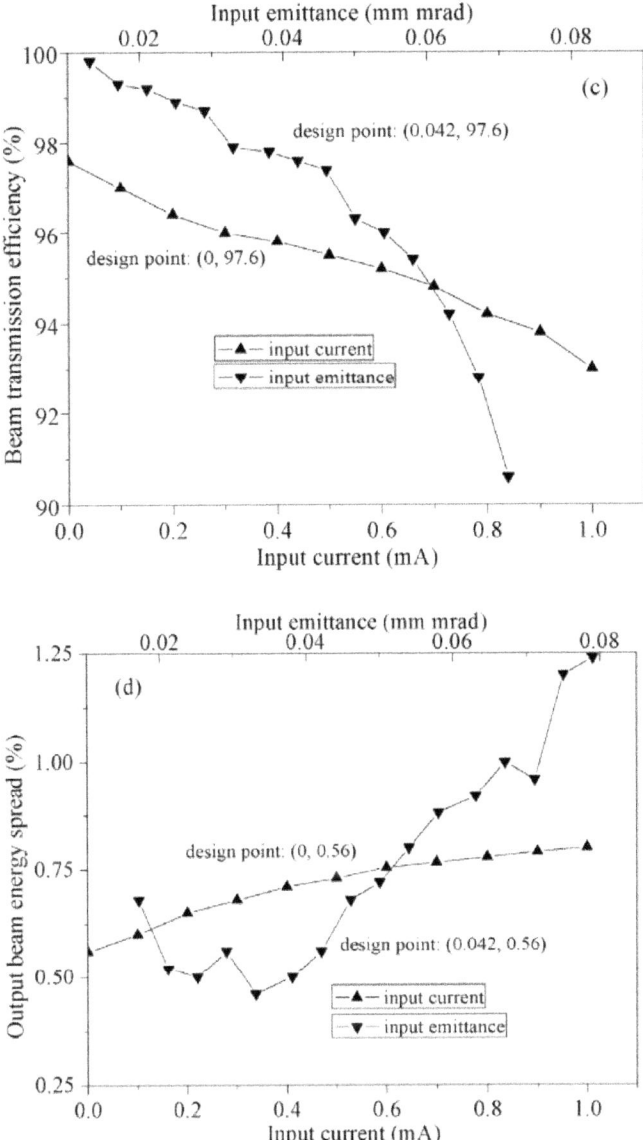

Figure 3: Stability studies of the beam dynamics design of the ladder IH-RFQ: beam transmission efficiency versus Twiss parameters in (a), output beam energy spread versus Twiss parameters in (b), beam transmission efficiency versus input current and emittance in (c), output beam energy spread versus input current and emittance in (d).

The design has been proved to be suitable for various ion species with different charge-tomass ratio $q/m0$ and even different frequencies, as long as the operating parameters are chosen appropriately. According to (3), it can be seen that the focusing strength B is proportional to $q/m0$, l^2 and V. Meanwhile, the length of one acceleration cell is b_v $\lambda /2$. For given electrodes structure, when $q/m0$ or λ changes, V and b_v i.e. the injection energy $W_i=1/2m_0v^2$ should be adjusted to keep B and the injection cell length unaltered, hence achieve the matching between the acceleration channel and the input beam. The beam dynamics parameters are listed in Table 2.

Table 2: Beam dynamics parameters of the ladder IH-RFQ

Ion species	$^{14}C^+$	$^{13}C^+$	$^{12}C^+$	$^4He^+$	$^4He^+$	$^4He^+$
Frequency f (MHz)	104	104	104	104	100	106.4
Inter-electrode voltage V (kV)	60	55.7	51.4	17.1	15.9	17.9
Injection energy W_i (keV)	40	37.1	34.3	11.4	10.6	12.0
Output energy W_{out} (keV)	500	464	428	143	132	150
Transmission T (%)	97	97	97	97	97	97
Energy spread (%)	0.6	0.6	0.7	1.0	1.1	1.0

RF Design of the Ladder IH-RFQ

According to the above beam dynamics design, the average electrode-tip radius is 3.1 mm corresponding to the average aperture radius 3.5 mm. Figure 4 shows the resonant model of CST Microwave Studio (MWS)[18] simulations for the ladder IH-RFQ. Simulation results of the electromagnetic field validated that the ladder IH-RFQ operates at $H_{21(0)}$ mode. As a consequence, the resonant frequency of it is unsurprisingly more than two times that of IHRFQ and about 1.3 times that of 4-Rod RFQ under the same transverse dimension, making it very attractive at frequencies above 100 MHz. Moreover, in respect that the electrodes are supported by completely symmetric stem structures, the quadrupole field of the ladder IHRFQ is ideal unlike the 4-Rod RFQ and IH-RFQ, the dipole components of which are typically 2-3% and 0.5%, respectively[19] .

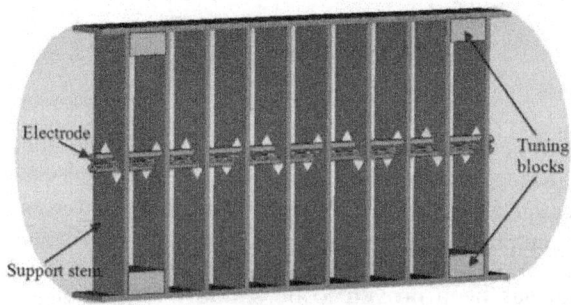

Figure 4: MWS simulation model of the ladder IH-RFQ.

Many authors have reported their studies on the RFQ cavity modeling[20,21]. The conventional electrical model are shown in Fig.5, consisting of the equivalent resistance, inductance, capacitance and parallel shunt impedance, denoted by R, L, C and R_p, respectively. According to the definitions of intrinsic quality factor $Q_0 = \omega W/P$ and shunt impedance $R_p = V^2/P$, with the resonant frequency $\omega = 2\pi f = 1/(LC)^{1/2}$, average energy stored in one period $W = 1/2 I^2 L = 1/2 CV^2$ (where I is peak current and V is peak voltage), and power dissipation $P = 1/2 I^2 R$, we have

$$Q_0 = \frac{1}{R}\sqrt{\frac{L}{C}} = \frac{R_P}{2}\sqrt{\frac{C}{L}}$$

(5)

For RFQ cavity having N basic resonant cells, the relation between the total R, L, C, I and the corresponding values of one cell R_{cell}, L_{cell}, C_{cell}, I_{cell} will be

$$I = NI_{cell} \, , \; C = NC_{cell} \, , \; L = L_{cell}/N \, , \; R = R_{cell}/N$$

(6)

when ignoring the end effect. Combining (5) with (6) we find

$$Q_0 = Q_{0,cell} \, , \; R_p = R_{p,cell}/N$$

(7)

where cell denote the values of single cell. For ladder IH-RFQ, Fig.6 shows the simplified circuit of one cell with

$$C_{cell} = 2C_{stem} + C_{rod} \; \colon \; L_{cell} = L_{stem}/2 \; \colon \; R_{cell} = R_{stem}/2$$

(8)

In (8) it is assumed that the resistance and inductance are mainly attributed to the stems, while the capacitance arises from the two pairs of opposite electrodes and adjacent stems with C_{rod} several times greater than C_{stem} generally.

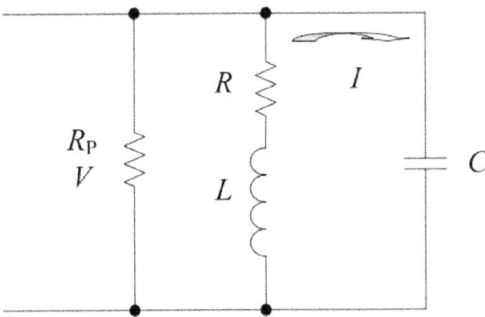

Figure 5: Equivalent circuit of RFQ cavity.

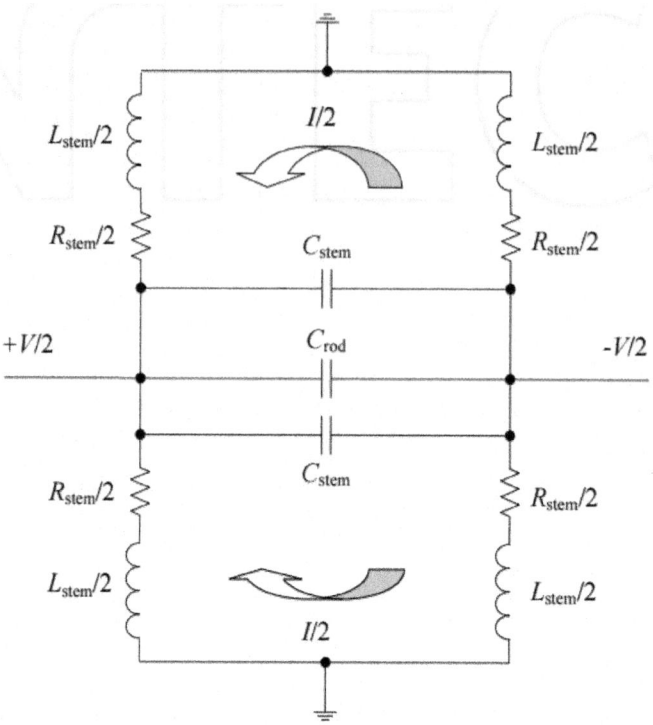

Figure 6: Equivalent circuit of single cell for the ladder IH-RFQ.

Based on the equivalent circuit analysis, table 3 lists the MWS-simulated rf properties of different RFQ structures at the same transverse size, which is in agreement with the theoretical predictions.

Table 3: Comparison of different RFQ structures at the same transverse size

RFQ Structures	Frequency (MHz)	Shunt impedance (kΩ)	Quality factor	The first High-mode Frequency (MHz)
IH-RFQ	46.5	273	5934	137.2
4-rod RFQ	77.9	155	5077	143.9
Ladder IH-RFQ	103.6	100	5192	156.3

Mode analysis of the RFQ was performed using the eigenmode solver of MWS. Fig. 7 shows the electric field, magnetic field and surface current distribution. It can be seen that the two parts of inductance and resistance of one stem are in parallel. These results agree with the electromagnetic field distribution of a H210 mode, and the expectation from Fig.6.

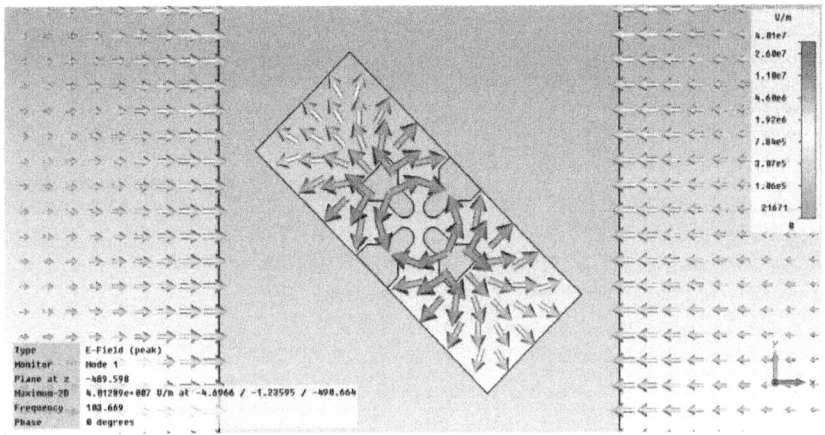

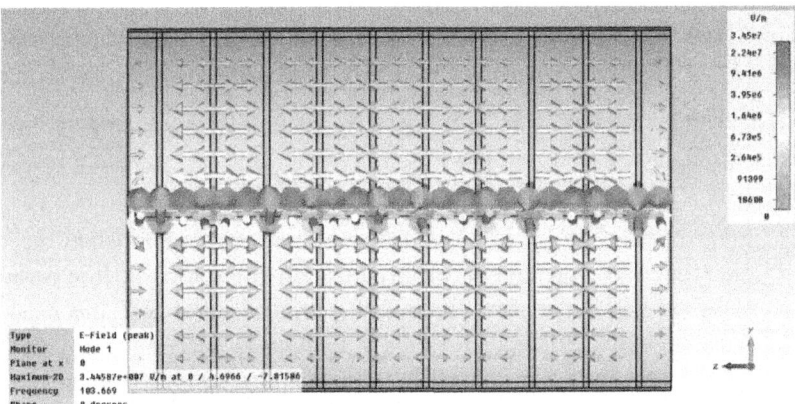

(a) Electric field among electrodes (up) and stems (bottom) of the ladder IH-RFQ

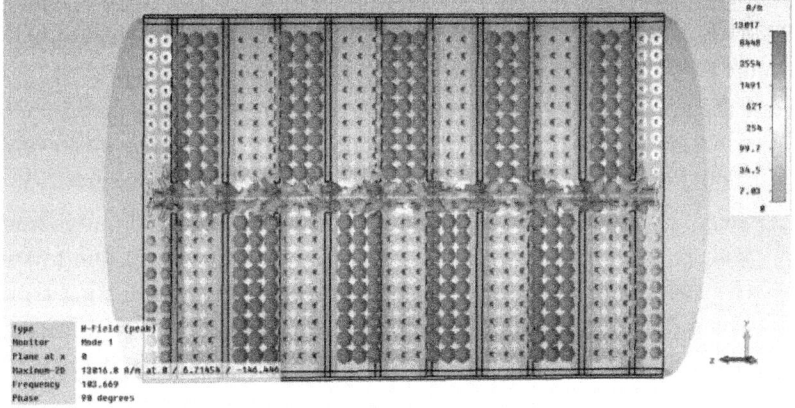

(b) Magnetic field distribution of the ladder IH-RFQ

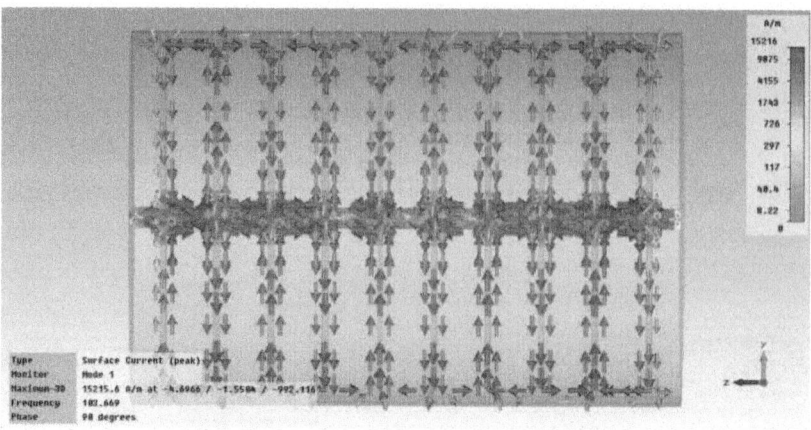

(c) Surface current flow of the ladder IH-RFQ

Figure 7: MWS-simulated results of the electromagnetic field and surface current flow of the ladder IH-RFQ.

The RFQ structure was optimized by parameter sweeping to find the maximum Q_0, R_p, and minimum P at 104 MHz. Some conclusive results are as follows.

- Stem number N_{stem}: The beam dynamics design established the inter-electrodes aperture and the electrode length, therefore the first parameter to be studied is the stem number N_{stem} used. Bigger N_{stem} means shorter distance between adjacent stems d, leading to decreasing L_{cell} and C_{cell}, but slightly increasing Rcell in view of more serious proximity effect. Taking (6) into account, R, L will decrease while C increase because of the increment in C_{stem}, finally resulting in bigger f as illustrated by MWS simulations. The cavity diameter D has to be scaled to keep f 104 MHz. Fig.8 shows the relevant simulation results, according to which 10 stems was employed.

- Cavity diameter D: As a variable D is always used to make f unaltered when sweeping the other parameters. When f needs to be lowered, D should be extended to make L_{stem}, C_{stem} as well as R_{stem} bigger.

- Stem spacing d: Over the 95 mm to 115 mm d range in 5 mm increments with N_{stem} =10, f decreases while Rp and Q0 increase due to smaller C_{stem}, R_{stem} and bigger L_{stem} with increasing d. After adjusting D, it was found that R_p and Q0 are approximately proportional to d, taking into account the end effects and the mechanical feasibility d=110 mm was eventually chosen.

- Stem width W_{stem}: The increase in C_{stem}, and decrease in L_{stem} as well as R_{stem} with W_{stem} over the simulation range result in increasing f and Q0 but decreasing Rp. By adjusting D, it could be seen that there is a peak value of R_p, this is why W_{stem} =130 mm was adopted to make power loss least.

The inter-electrodes voltage is usually supposed to be constant longitudinally when carrying out the beam dynamics design. So it has long been an important topic that how to tune the voltage unflatness of the RFQ owing to the end effect and modulated electrodes to ensure a good beam quality. For the ladder IH-RFQ, adjustable tuning plates as shown in Fig.4 can be inserted into two neighboring stems symmetrically up and down, which is similar to the case of the 4-Rod RFQ[22]. The tuning plates serve as short pieces of stems, changing their inductances thus the frequency and field strength. To verify their tuning capability, corresponding simulations were performed for different tuning plate heights 0, 50 and 100 mm, of which the results are plotted in Fig.9. Here electric field is equal to the voltage between relevant electrodes, since the electrodes are unmodulated with uniform aperture radius. It can be seen that the field unflatness achieves its optimum value 2% when the tuning height is 100 mm. Tuning plates cause increment in the eigenfrequency of RFQ, so f is generally set to be several percents lower than the desired value to leave enough tuning space.

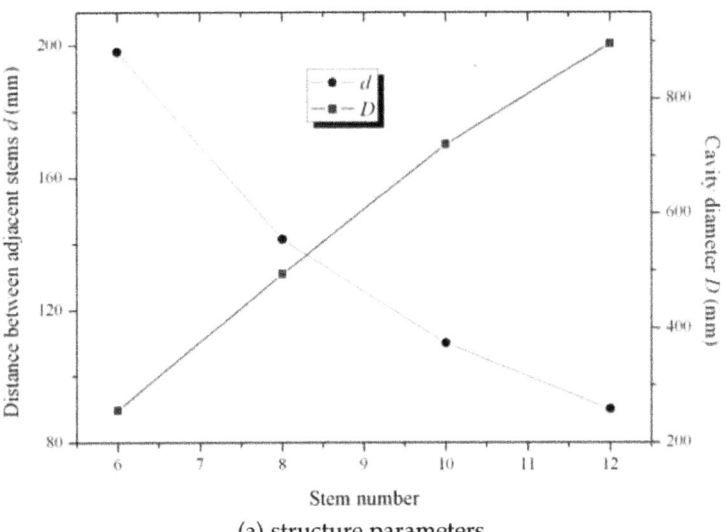

(a) structure parameters

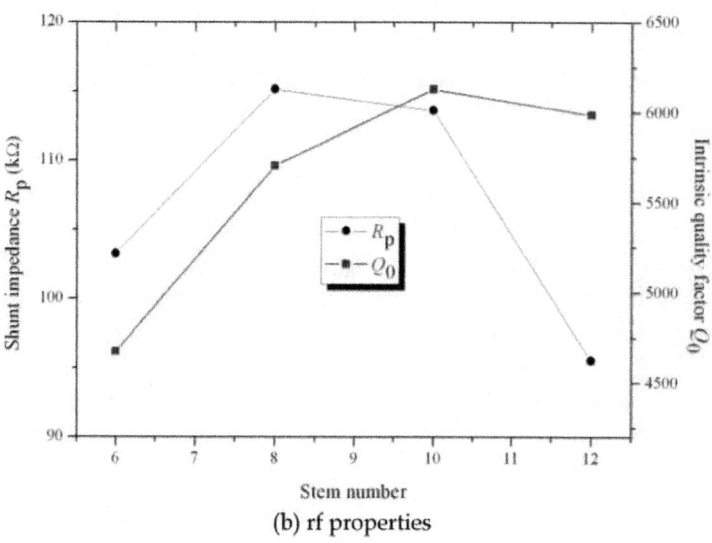

(b) rf properties

Figure 8: Variations of the ladder IH-RFQ parameters with stem number.

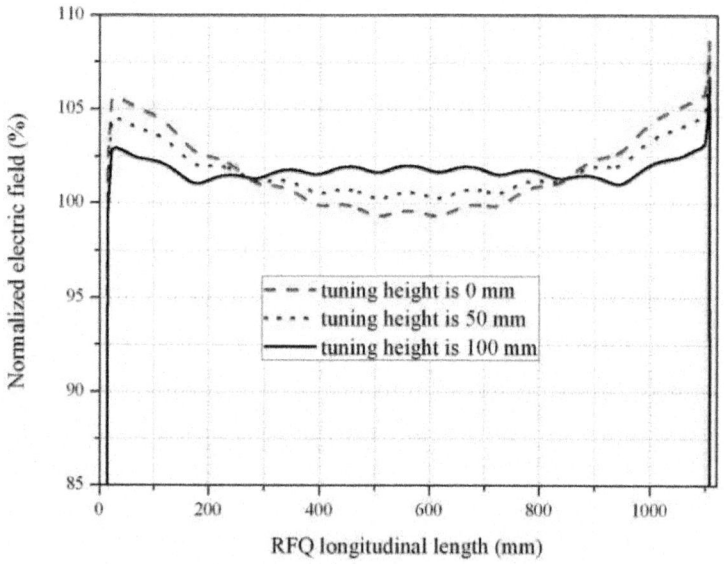

RFQ longitudinal length (mm)

Figure 9: Variations of the field unflatness with tuning plates height of the ladder IH-RFQ.

Experimental Studies on the Ladder IH-RFQ

The constructed ladder IH-RFQ cavity is shown in Fig.10. The geometrical parameters are as follows: the cavity is 720 mm in diameter and 1123 mm in length; 10 supporting stems with width and spacing of 130 mm and 110 mm, respectively, are used. The stems are mounted on two fixing plates up and down to improve the mechanical performance and in the interest of installation. The stems and plates are made of oxygen free copper, and there are water-cooling channels in them. The inner wall of the cavity tank is well plated with copper of more than 0.1 mm in thickness to improve the rf performances. In order to fix the stems and plates on the tank, four copper blocks with tuning thickness 30 mm are used, which will make the resonant frequency higher and improve the field unflatness at the same time.

Figure 10: The constructed ladder IH-RFQ cavity.

On the basis of transient simulations with MWS, a magnetic coupling loop with size of about 90 mm´50 mm in between two neighboring stems was established for rf power feeding of the ladder IH-RFQ. The critical coupling condition was met after proper adjustment of the loop orientation, and the result is shown in Fig.11(a). The cold measurement was performed by means of a vector network analyzer and the capacitance perturbation method. S-parameters were measured using an Angilent 8753E network analyzer, and the loaded quality factor Q_L was obtained by 3 dB method. Consequently, the unloaded or intrinsic quality factor Q_0 could be computed using $Q_0 = Q_L(1+b_c)$, where the coupling factor b_c is defined as Q_0/Q_e with the external quality factor Q_e. During the experiment, a very small pick-up loop was used to weaken the influence of it on the cavity properties, as a result, $b_c \approx 1$ for critical coupling, i.e. $Q_0 = Q_e$. Referring to the shunt impedance, the perturbation capacitor method was adopted as described in Ref.[23,24]. According to the definitions of intrinsic quality factor $Q_0 = \omega W/P$ and shunt impedance $R_p = V^2/P$, we have

$$R_p = \frac{2Q_0}{\omega C}$$

(9)

When using a perturbation capacitance ΔC, the perturbed resonant frequency of the cavity f_{per} becomes

$$f_{per} = \frac{1}{2\pi\sqrt{L(C + \Delta C)}}$$

(10)

So the effective capacitance of the cavity can be thus obtained by

$$C = \frac{\Delta C}{\left(f / f_{per}\right)^2 - 1}$$

(11)

Finally R_p is calculated using (9), and the square root of it is treated as the relative value of the inter-electrode voltage. The locations of the perturbation capacitor are moved from stem to stem as shown in Fig.11(b), in the interest of the average Rp and voltage unflatness. Figure 12 is the adjustment result of the feeding point, and Fig.13 shows the S_{21} parameter with the loaded quality factor 1861 at 106.7 MHz. The cold measurement result of the shunt impedance is 102 kΩ compared with the simulation value 110 kΩ. The measured interelectrode voltage unflatness and the simulated result are compared in Fig.14.

Finally, the high-power test and beam test of the ladder IH-RFQ were performed[25]. The rf and measurement system of the experiment is plotted in Fig.15. We used one pulse generator to control the ECR source[26] and the rf transmitter at the same time. The rf power is provided by the FM Transmitter, 816R-4, operating in the 88~108 MHz frequency range with the maximum rf output of 27.5 kW, the excitation of which is from a CE 802A solidstate FM wideband exciter. The input signal of the exciter is given by a signal modulation system consisting of a CW signal source, the pulse generator, a high frequency (HF) switch and a 30 dB solid-state power amplifier. The output of the transmitter is applied to the RFQ cavity via a low-pass filter and directional coupler. The beam current was measured using Faraday cups with the help of standard pick-up resistances.

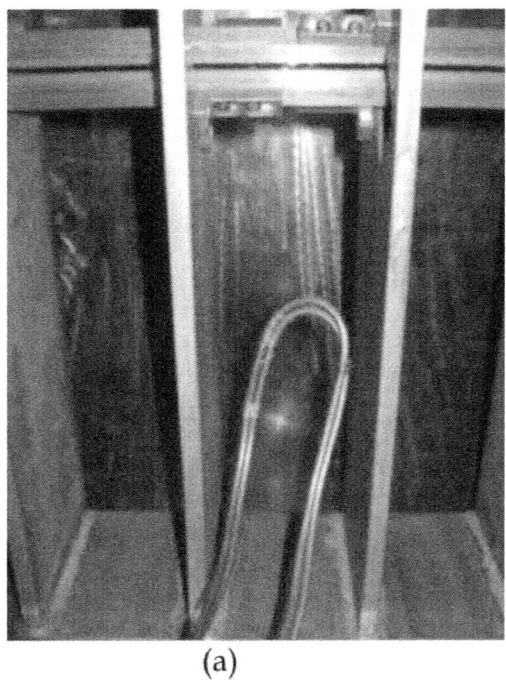

(a)

(b)

Figure 11: The critical coupling loop (a) and the perturbation capacitor (b) of the ladder IHRFQ.

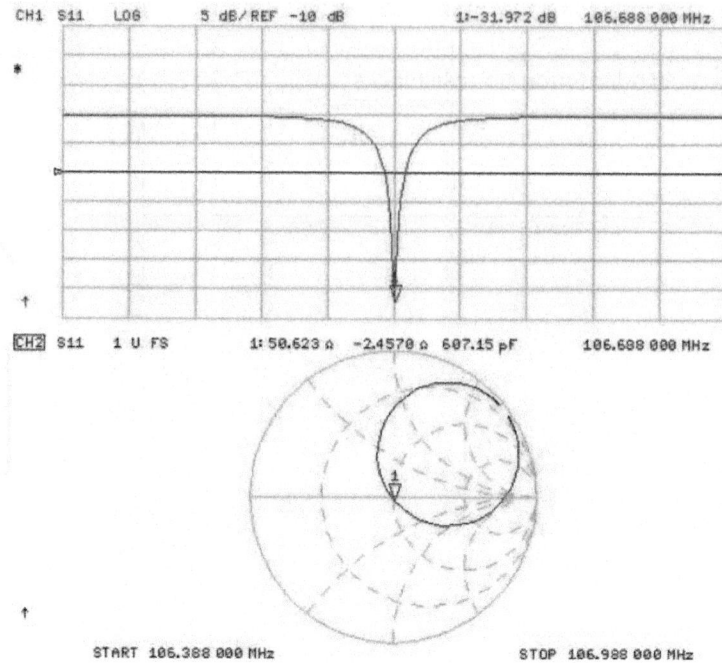

Figure 12: The critical coupling point of the ladder IH-RFQ: the S11 parameter (up) and Smith Chart (down).

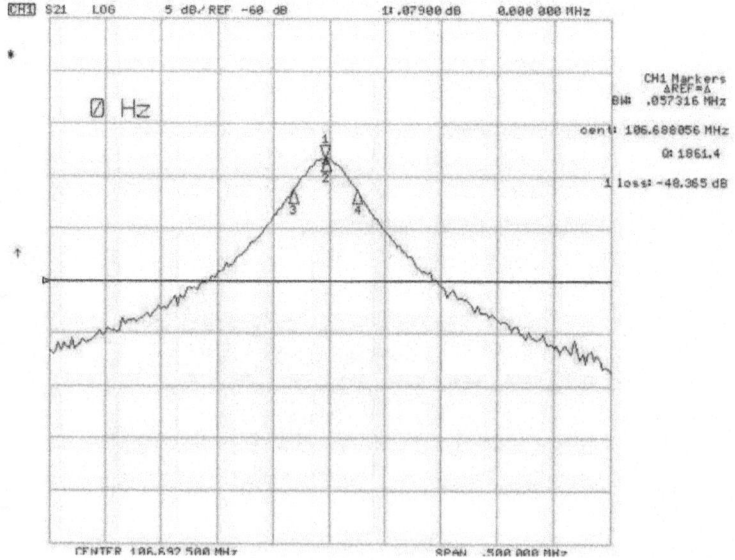

Figure 13: The measured S21 parameter of the ladder IH-RFQ.

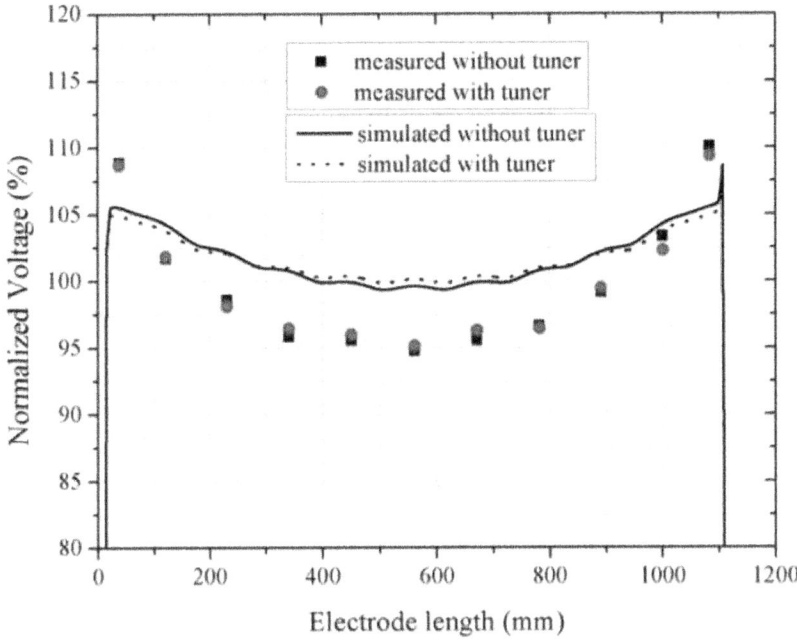

Figure 14: Simulated and measured voltage unflatness of the ladder IH-RFQ.

Commissioning of the transmitter was carried out at first, because it had been more than 10 years since the last operation. During the experiment, a water load with 50 Ω impedance was used to consume the rf power from the transmitter. The inter-electrode voltage was started to be measured when the high vacuum in the RFQ cavity reached 10^{-5} Pa with both cooling water and high rf power. To obtain the inter-electrode voltage, a high purity Ge detector cooled by liquid nitrogen was used to measure the energy spectrum of X-ray [8]. An ORTEC computer multi-channel system was employed to deal with the signal, which consists of a preamplifier, a master amplifier, a PCI computer multi-channel card and anotebook PC. The maximum electron energy in the measured Roentgen ray spectrum corresponds to the inter-electrode voltage. The system was calibrated using two standard radiation source [241]Am and [152]Eu. After adjusting the amplification factor properly, g ray of [241]Am 59.54 keV locates at channel 316 while [152]Eu 344.31 keV at channel 1889, then two peaks appear at channel 661 and channel 1340 corresponding to [152]Eu 121.78 keV and [152]Eu 244.66 keV, respectively. The above 4 points fit a well linearity, which means the calibration is successful.

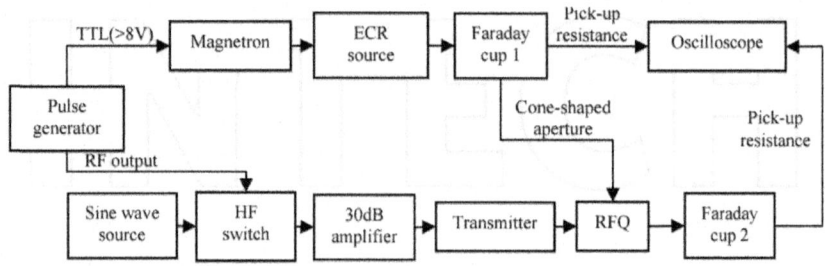

Figure 15: The rf and measurement system of the experiment for the ladder IH-RFQ.

From the measured Roentgen spectrum at rf power 26.4 kW as shown in Fig.16, it can be found that the inter-electrode voltage reaches 72.1 kV, which is higher than the required value 60 kV. What should be paid attention to is that the spectrum was measured at the glass window locating near the entrance of the RFQ. According to Fig.14, the inter-electrode voltage unflatness is about ±7%, as a result, the average value of it along the cavity longitudinally should be 72.1/1.07 kV, i.e. 67.3 kV in this power case. All the measured voltages are normalized in this way. Figure 17 plots the variation of the square of the interelectrode voltage with rf power, from which it can be seen that the linearity is well. The specific shunt impedance is defined as $R_s = (V^2/P) \cdot L = R_p \cdot L$, where L=1123 mm is the length of the RFQ. The slope of the fitting line that means the shunt impedance is 162 kΩ, so the corresponding specific shunt impedance is 178 kΩïm. The result is better than that of simulation and cold measurement, which is because that the conductivity was set to be 5.0×10^7 S/m not 5.9×10^7 S/m in MWS simulations and the rf properties were improved via advance power test. The temperature of cooling water increased from 15.7 °C to 23.4 °C while the rf power increasing from 0 to 29.7 kW, and no sparking phenomenon appeared during the experiment. The results above indicate that the RFQ cavity possesses well mechanical strength and is water-cooled effectively.

The beam test has been performed using $^4He^+$ beam. The output current and transmission efficiency of the ladder IH-RFQ varying with the rf power fed into it are shown in Fig. 18 when the extraction current from the ECR source is 30 mA. The maximum transmission efficiency reaches 70%. During the experiment, the highest output beam is 210 mA, and the transmission efficiency is better than 70% at current of several tens mA. It should be figured out that the transmission efficiency is mainly limited by the unmatched beam injected to the RFQ, especially for high current beam. Further efforts will be made to improve it. The following works will also include the beam test of heavier ions, measurement of the energy spectrum, and some application studies on the RFQ as an implanter.

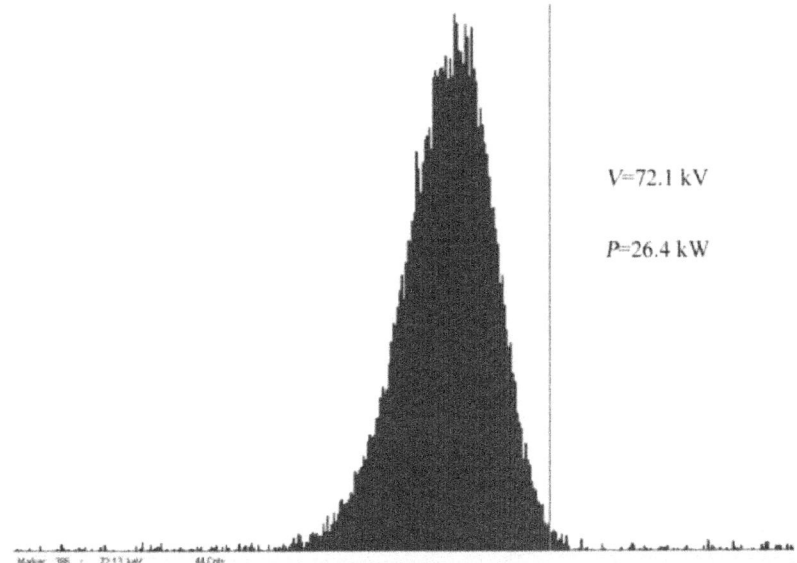

Figure 16: Roentgen spectrum at 26.4kW rf power.

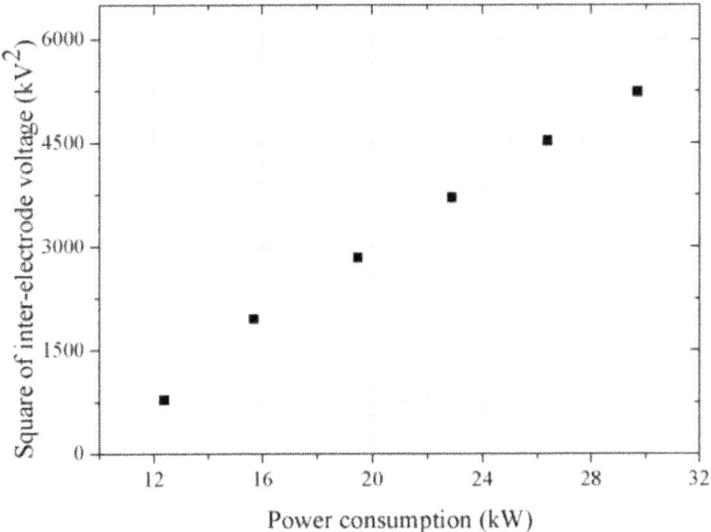

Figure 17: Square of the inter-electrode voltage vs rf power.

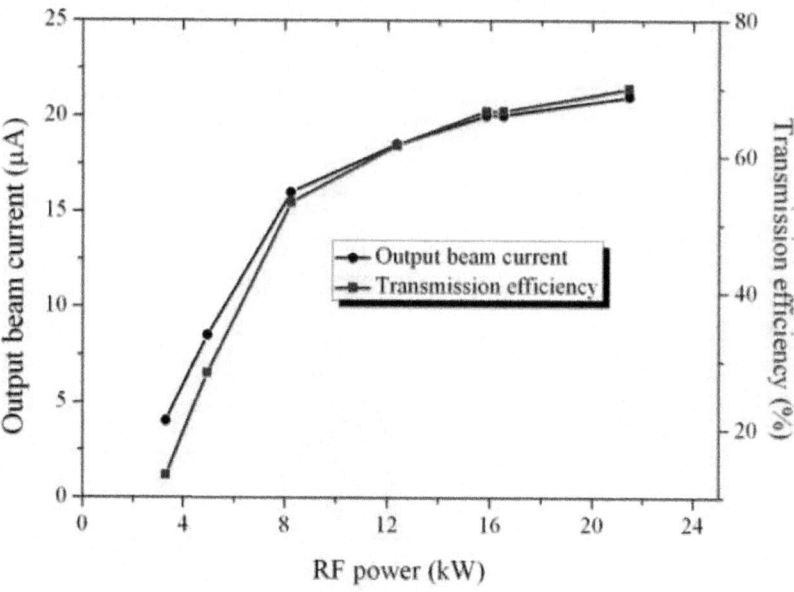

Figure 18: The output current and transmission efficiency of the RFQ vs rf power at the extraction current 30mA(^{4}He$^+$).

SUMMARY

In conclusion, the RFQ accelerator is a competent ion implanter with many advantages such as high energy (MeV), high current (mA-mA), variable energy and ion species, compact size and so on. Further studies will be made on the practical usage of the RFQs in the ion implantation field. As for the ladder IH-RFQ accelerator of Peking University, after the further beam test is finished, some ion injection experiment will be made for the surface modification of materials.

ACKNOWLEDGEMENT

The authors would like to thank Professor Dr. U. Ratzinger of IAP, J. W. Goethe University Frankfurt am Main, for the helpful discussions. Y. C. Nie would like to thank the China Scholarship Council (CSC) for the financial supports when he was in Frankfurt. This work was supported by the National Natural Science Foundation of China (Grant No. 10775009 and 11079001).

REFERENCES

1. M. Kapchinsky, V. A. Teplyakov, Linear Ion Accelerator with Spatially Homogeneous Strong Focusing, Prib. Tekh. Eksp., 1970, No.2, p.19.

2. T. P. Wangler, Principles of RF Linear Accelerators, Wiley Series in Beam Physics and Accelerator Technology, 1998, p.225-257.

3. H. Klein, Development of the Different RFQ Accelerating Structures and Operation Experience, IEEE Trans. Nucl. Sci., 1983, Vol. NS-30, p.3313.

4. L. Young, 25 Years of Technical Advances in RFQ Accelerators, Proc. of PAC2003, p.60.

5. A. Schempp, Research application of RFQ linac, Nuclear Instruments and Methods in Physics Research Section B, 1995, Vol.99, p.688-693.

6. Ren Xiao-Tang, Lu Yuan-Rong, Yu Jin-Xiang, et al., Experimental research on simultaneous acceleration of positive and negative ions with equal q/m in an ISR RFQ, High Energy Physics and Nuclear Physics (in Chinese), 2000, Vol.24, p.347- 351.

7. U. Ratzinger, et al., The GSI 36 MHz high-current IH-type RFQ and HIIF-relevant extensions, Nuclear Instruments and Methods in Physics Research Section A, 1998, Vol.415, p.281-286.

8. Y. R. Lu, C. E. Chen, J. X. Fang, et al., Investigation of high duty factor ISR RFQ-1000, Nuclear Instruments and Methods in Physics Research Section A, 2003, Vol.515, p.394-401.

9. Kensuke Amemiya, Junya Ito and Katsumi Tokiguchi, Aluminum ion implantation using a variable energy RFQ implanter, Proc. of EPAC1998, p.2419.

10. Byung-Wuek Lee, Jong-Won Kim and Holger Podlech, Design study of an RFQ for high-energy ion implantation, Journal of the Korean Physical Society, 2006, Vol.48, p.810-814.

11. A. Schempp, et al., High duty factor nitrogen RFQ as a prototype high current implanter, Nuclear Instruments and Methods in Physics Research Section B, 1998, Vol.139, p.411-417.

12. Zhiyu Guo, et al., Feasibility studies of RFQ based 14C accelerator mass spectrometry, Nuclear Instruments and Methods in Physics Research Section B, 2007, Vol.259, p.204-207.

13. NIE Yuan-Cun, LU Yuan-Rong, CHEN Jia-Er, et al., Theoretical design of a 104 MHz ladder type IH-RFQ accelerator, Chinese Physics Letters, 2010, Vol.27, p.112901.

14. Fang Jiaxun, et al., RFQ accelerators for ion implantation, Nuclear Techniques (in Chinese), 2006, Vol.29, p.140-144.

15. J. W. Staples, Reducing RFQ Longitudinal Emittance, Proc. of LINAC1994, p. 755.

16. K. R. Crandall, R. H. Stokes, T. P. Wangler, RF quadrupole beam dynamics design studies, Proc. of LINAC1979, BNL-51134, p.205.

17. X. Q. Yan, et al., Matched and equipartitioned design method for modern high-intensity radio frequency quadrupole accelerators, Nuclear Instruments and Methods in Physics Research Section A, 2007, Vol.577, p.402-408.

18. www.cst.com

19. H. Podlech, et al., Electromagnetic design of an 80.5 MHz RFQ for the RIA driver Linac, Proc. of EPAC2002, p.942-944.

20. R. M. Hutcheon, An equivalent circuit model of the general 3-dimensional RFQ, IEEE Trans. Nucl. Sci., 1983, Vol. NS-30, p.3524-3526.

21. J. X. Fang, and A. Schempp, Equivalent circuit of a 4-Rod RFQ, Proc. of EPAC92, p.1331- 1333.

22. J. Schmidt, et al., Tuning of the 4-Rod RFQ for MSU, Proc. of IPAC'10, p.762-764.

23. P. Fischer and A. Schempp, Tuning of a 4-rod cw-mode RFQ accelerator, Proc. of LINAC06, p.728-730.

24. P. G. Bricault, Methods for measurement of the specific shunt impedance of RFQ, TRIUMF DESIGN NOTE TRI-DN-94-30, 1994.

25. Lu Yuan-Rong, Chen Wei, Nie Yuan-Cun, et al., Power test of the ladder IH-RFQ accelerator at Peking University, Chinese Physics Letters, 2011, Vol.28(7), p.072901.

26. M. Zhang, S. X. Peng, Z. Z. Song, et al., Experimental results of an ECR oxygen source and a LEBT system for 1 MeV ISR RFQ accelerator upgrade project, Chinese Physics C, 2008, Vol.32(Suppl.I), p.220-222.

Chapter 9

GEDANKEN EXPERIMENT FOR REFINING THE UNRUH METRIC TENSOR UNCERTAINTY PRINCIPLE VIA SCHWARTZ SHIELD GEOMETRY AND PLANCKIAN SPACE-TIME WITH INITIAL NONZERO ENTROPY AND APPLYING THE RIEMANNIAN-PENROSE INEQUALITY AND INITIAL KINETIC ENERGY FOR A LOWER BOUND TO GRAVITON MASS (MASSIVE GRAVITY)

Andrew Walcott Beckwith

College of Physics, Chongqing University Huxi Campus, Chongqing, China

ABSTRACT

This paper is with the permission of Stepan Moskaliuk similar to what he will put in the conference proceedings of the summer teaching school and workshop for Ukrainian PhD physics students as given in Bratislava, as of summer 2015. With his permission, this paper will be in part reproduced here for this journal. First of all, we restate a proof of a highly localized special case of a metric tensor uncertainty principle first written up by Unruh. Unruh did not use the Roberson- Walker geometry which we do, and it so happens that the dominant metric tensor we will be examining, is variation in δg_u. The metric tensor variations given by δg_{rr}, $\delta g_{\theta\theta}$ and $\delta g_{\phi\phi}$ are negligible, as compared to the variation δg_u. Afterwards, what is referred to by Barbour as emergent duration of time δt is from the Heisenberg Uncertainty principle (HUP) applied to δg_u in such a way as to give, in the Planckian space-time regime a nonzero minimum non zero lower ground to a massive graviton, $m_{graviton}$. The lower bound to the massive graviton is influenced by δg_u and kinetic energy which is in the Planckian emergent duration of time δt as $(E - V)$. We find from δg_u version of the Heisenberg Uncertainty Principle (HUP), that the quantum value of the $\Delta t \cdot \Delta E$ Heisenberg Uncertainty Principle (HUP) is likely not recoverable due to $\delta g_{tt} \neq O(1) \sim g_{tt} \equiv 1$. i.e. $\delta g_{tt} \neq O(1)$ is consistent with non-curved space, so $\Delta t \cdot \Delta E \geq \hbar$ no longer holds. This even if we take the stress energy tensor approximation $T_{ii} = \mathbf{diag}(\rho, -p, -p, -p)$ where the fluid approximation is used. Our treatment of the inflaton is via Handley et al., where we consider the lower mass limits of the graviton as due to when the inflaton is many times larger than a Potential energy, with a kinetic energy (KE) proportional to $\rho_w \propto a^{-3(1-w)} \sim g^{*} T^{4}$, with g^{*} initial degrees of freedom, and T initial temperature. Leading to non-zero initial entropy as stated in Appendix A. In addition we also examine a Ricci scalar value at the boundary between Pre Planckian to Planckian regime of space-time, setting the magnitude of k as approaching flat space conditions right after the Planck regime. Furthermore, we have an approximation as to initial entropy production. $N \sim S_{\text{initial(graviton)}} \sim 10^{37}$ Finally, this entropy is N, and we get an initial version of the cosmological "constant" as Appendix D which is linked to initial value of a graviton mass. Appendix E is for the Riemannian-Penrose inequality, which is either a nonzero NLED scale factor or quantum bounce as of LQG. Note that, Appendix F gives conditions so that a pre Planckian kinetic energy (inflaton) value greater than Potential energy occurs, which is foundational to the lower bound to Graviton mass. We will in the future add more structure to this calculation so as to confirm via a precise calculation that the lower bound to the graviton mass, is about 10^{-70} grams. Our lower bound is a dimensional approximation so far. We will make it exact. We conclude in this document with Appendix G, which is comparing our Pre Planckian space-time metric Heisenberg Uncertainty Principle with the generalized uncertainty principle in quantum gravity. Our result is different from the one given by Ali, Khali and Vagenas, in which our energy fluctuation is not proportional to that of processes of energy connected to Black hole physics, and we also allow for the possibility of Pre Planckian time. Whereas their result (and the generalized string theory Heisenberg Uncertainty principle) have a more limited regime of interpolation of final results. We do come up with equivalent bounds to recover $\delta g_{tt} \sim \textbf{small-value} \neq O(1)$ and the deviation of fluctuations of energy, but with very specific bounds upon the parameters of Ali, Khali, and Vagenas, but this has to be more fully explored. Finally, we close with a comparison of what this new Metric tensor uncertainty principle presages as far as avoiding the Bicep 2 mistake, and the different theories of gravity, as reviewed in Appendix H.

INTRODUCTION

The first matter of business will be to introduce a framework of the speed of gravitons in "heavy gravity". Heavy Gravity is the situation where a graviton has a small rest mass and is not a zero mass particle, and this existence of "heavy gravity" is important since eventually, as illustrated by Will [1] [2] gravitons having a small mass could possibly be observed via their macroscopic effects upon astrophysical events. Secondly, our manuscript's inquiry also will involve an upper bound to the rest mass of a graviton. The second aspect of the inquiry of our manuscript will be to come up with a variant of the Heisenberg Uncertainty principle (HUP), involving a metric tensor, as well as the Stress energy tensor, which will in time allow us to establish a lower bound to the mass of a graviton, preferably at the start of cosmological evolution. The article concludes in its last section as to why a statement by Mukhanov in Marcel Grossman 14, 2015, Rome, that a multiverse contribution to a new universe would have a causal barrier averaging of time contributions even if there were contributions from a multiverse, so there was only one space-time contribution is possibly indefensible.

We reference what was done by Will in his living reviews of relativity article as to the "Confrontation between GR and experiment". Specifically we make use of his experimentally based formula of [1] [2] , with v_{graviton} the speed of a graviton, and m_{graviton} the rest mass of a graviton, and E_{graviton} in the inertial rest frame given as:

$$\left(\frac{v_{\text{graviton}}}{c}\right)^2 = 1 - \frac{m_{\text{graviton}}^2 c^4}{E_{\text{graviton}}^2}$$

(1)

Furthermore, using [2] , if the rest mass of a graviton is very small we can make a clear statement of

$$\frac{v_{\text{graviton}}}{c} = 1 - 5 \times 10^{-17} \cdot \left(\frac{200\,\text{Mpc}}{D}\right) \cdot \left(\frac{\Delta t}{1\,\text{sec}}\right)$$

$$\doteq 1 - 5 \times 10^{-17} \cdot \left(\frac{200\,\text{Mpc}}{D}\right) \cdot \left(\frac{\Delta t = \Delta t_a - (1+z) \cdot \Delta t_b}{1\,\text{sec}}\right)$$

$$\Leftrightarrow \frac{2m_{\text{graviton}} c^2}{E_{\text{graviton}}} \approx 5 \times 10^{-17} \cdot \left(\frac{200\,\text{Mpc}}{D}\right) \cdot \left(\frac{\Delta t_a - (1+z) \cdot \Delta t_b}{1\,\text{sec}}\right)$$

(2)

Here, Δt_a is the difference in arrival time, and Δt_e is the difference in emission time/in the case of the early Universe, i.e. near the big bang, then if in the beginning of time, one has, if we assume that there is an average

$E_{\text{graviton}} \approx \hbar \cdot \omega_{\text{graviton}}$, and

$$\Delta t_a \sim 4.3 \times 10^{17} \text{ sec}$$

$$\Delta t_e \sim 10^{-33} \text{ sec}$$

$$z \sim 10^{50} \tag{3}$$

Then, $\left(\dfrac{\Delta t_a - (1+z)\cdot \Delta t_b}{1 \text{ sec}}\right) \sim 1$ and if $D \sim 4.6 \times 10^{26} \text{ meters} = \text{radii(universe)}$, so one can set

$$\left(\frac{200 \text{ Mpc}}{D}\right) \sim 10^{-2} \tag{4}$$

And if one sets the mass of a graviton [3] into Equation (1), then we have in the present era, that if we look at primordial time generated gravitons, that if one uses the

$$\Delta t_a \sim 4.3 \times 10^{17} \text{ sec}$$

$$\Delta t_e \sim 10^{-33} \text{ sec}$$

$$z \sim 10^{55} \tag{5}$$

Note that the above frequency, for the graviton is for the present era, but that it starts assuming genesis from an initial inflationary starting point which is not a space-time singularity.

Note this comes from a scale factor, if $z \sim 10^{55} \Leftrightarrow a_{\text{scale-factor}} \sim 10^{-55}$, i.e. 55 orders of magnitude smaller than what would normally consider, but here note that the scale factor is not zero, so we do not have a space-time singularity.

We will next discuss the implications of this point in the next section, of a non-zero smallest scale factor. Secondly the fact we are working with a massive graviton, as given will be given some credence as to when we obtain a lower bound, as will come up in our derivation of modification of the values [3]

$$\left\langle (\delta g_{uv})^2 (\hat{T}_{uv})^2 \right\rangle \geq \frac{\hbar^2}{V_{\text{Volume}}^2} \xrightarrow{uv \to tt} \left\langle (\delta g_{tt})^2 (\hat{T}_{tt})^2 \right\rangle \geq \frac{\hbar^2}{V_{\text{Volume}}^2}$$

$$\& \quad \delta g_{rr} \sim \delta g_{\theta\theta} \sim \delta g_{\phi\phi} \sim 0^+ \tag{6}$$

The reasons for saying this set of values for the variation of the other metric components will be in the 3rd section and it is due to the smallness of the square of the scale factor in the vicinity of Planck time interval.

NON ZERO SCALE FACTOR, INITIALLY AND WHAT THIS IS TELLING US PHYSICALLY. STARTING WITH A CONFIGURATION FROM UNRUH

Begin with the starting point of [4] [5]

$$\Delta l \cdot \Delta p \geq \frac{\hbar}{2} \tag{7}$$

We will be using the approximation given by Unruh [4] [5] , of a generalization we will write as

$$\left(\Delta l\right)_{ij} = \frac{\delta g_{ij}}{g_{ij}} \cdot \frac{l}{2}$$

$$\left(\Delta p\right)_{ij} = \Delta T_{ij} \cdot \delta t \cdot \Delta A \tag{8}$$

If we use the following, from the Roberson-Walker metric [6] .

$$g_{tt} = 1$$
$$g_{rr} = \frac{-a^2(t)}{1 - k \cdot r^2}$$
$$g_{\theta\theta} = -a^2(t) \cdot r^2$$
$$g_{\phi\phi} = -a^2(t) \cdot \sin^2\theta \cdot d\phi^2 \tag{9}$$

Following Unruh [4] [5] , write then, an uncertainty of metric tensor as, with the following inputs

$$a^2(t) \sim 10^{-110}, r \equiv l_p \sim 10^{-35} \text{ meters} \tag{10}$$

Then, the surviving version of Equation (7) and Equation (8) is, then, if $\Delta T_{tt} \sim \Delta\rho$

$$V^{(4)} = \delta t \cdot \Delta A \cdot r$$

$$\delta g_{tt} \cdot \Delta T_{tt} \cdot \delta t \cdot \Delta A \cdot \frac{r}{2} \geq \frac{\hbar}{2}$$

$$\Leftrightarrow \delta g_{tt} \cdot \Delta T_{tt} \geq \frac{\hbar}{V^{(4)}} \tag{11}$$

This Equation (11) is such that we can extract, up to a point the HUP principle for uncertainty in time and energy, with one very large caveat added, namely if we use the fluid approximation of space-time [6] for the stress energy tensor as given in Equation (12) below.

$$T_{ii} = \text{diag}\left(\rho, -p, -p, -p\right) \tag{12}$$

Then

$$\Delta T_{tt} \sim \Delta\rho \sim \frac{\Delta E}{V^{(3)}}$$

(13)

Then, Equations (11)-(13) together yield

$$\delta t \Delta E \geq \frac{\hbar}{\delta g_{tt}} \neq \frac{\hbar}{2}$$

Unless $\delta g_{tt} \sim O(1)$

(14)

How likely is $\delta g_{tt} \sim O(1)$? Not going to happen. Why? The homogeneity of the early universe will keep

$$\delta g_{tt} \neq g_{tt} = 1$$

(15)

In fact, we have that from Giovannini [6] , that if ϕ is a scalar function, and $a^2(t) \sim 10^{-110}$, then if

$$\delta g_{tt} \sim a^2(t) \cdot \phi \ll 1$$

(16)

Then, there is no way that Equation (14) is going to come close to $\delta t \Delta E \geq \frac{\hbar}{2}$. Hence, the Mukhanov sugges-

tion as will be discussed toward the end of this article, is not feasible. Finally, we will discuss a lower bound to the mass of the graviton.

HOW WE CAN JUSTIFYING WRITING VERY SMALL $\delta g_{rr} \sim \delta g_{\theta\theta} \sim \delta g_{\phi\phi} \sim 0^+$ VALUES

To begin this process, we will break it down into the following co ordinates

In the rr, $\theta\theta$ and $\phi\phi$ coordinates, we will use the Fluid approximation, $T_{ii} = \text{diag}(\rho, -p, -p, -p)$ [7] with

$$\delta g_{rr} T_{rr} \geq -\left| \frac{h \cdot a^2(t) \cdot r^2}{V^{(4)}} \right| \xrightarrow{a \to 0} 0$$

$$\delta g_{\theta\theta} T_{\theta\theta} \geq -\left| \frac{h \cdot a^2(t)}{V^{(4)}(1 - k \cdot r^2)} \right| \xrightarrow{a \to 0} 0$$

$$\delta g_{\phi\phi} T_{\phi\phi} \geq -\left| \frac{h \cdot a^2(t) \cdot \sin^2\theta \cdot d\phi^2}{V^{(4)}} \right| \xrightarrow{a \to 0} 0$$

(17)

If as an example, we have negative pressure, with T_{rr}, $T_{\theta\theta}$ and $T_{\phi\phi} < 0$, and $p = -\rho$, then the only choice we have, then is to set $\delta g_{rr} \sim \delta g_{\theta\theta} \sim \delta g_{\phi\phi} \sim 0^+$, since there is no way that $p = -\rho$ is zero valued.

Having said this, the value of δg_{tt} being non zero, will be part of how we will be looking at a lower bound to the graviton mass which is not zero.

LOWER BOUND TO THE GRAVITON MASS USING BARBOUR'S EMERGENT TIME

In order to start this approximation, we will be using Barbour's value of emergent time [8] [9] restricted to the Plank spatial interval and massive gravitons, with a massive graviton [10]

$$\left(\delta t\right)^2_{emergent} = \frac{\sum_i m_i l_i \cdot l_i}{2\cdot\left(E-V\right)} \rightarrow \frac{m_{graviton} l_P \cdot l_P}{2\cdot\left(E-V\right)}$$

(18)

Initially, as postulated by Babour [8] [9] , this set of masses, given in the emergent time structure could be for say the planetary masses of each contribution of the solar system. Our identification is to have an initial mass value, at the start of creation, for an individual graviton.

If $\left(\delta t\right)^2_{emergent} = \delta t^2$ in Equation (11), using Equation (11) and Equation (18) we can arrive at the identification of

$$m_{graviton} \geq \frac{2\hbar^2}{\left(\delta g_{tt}\right)^2 l_P^2} \cdot \frac{\left(E-V\right)}{\Delta T_{tt}^2}$$

(19)

Key to Equation (19) will be identification of the kinetic energy which is written as $E-V$. This identification will be the key point raised in this manuscript. Note that [11 raises the distinct possibility of an initial state, just before the "big bang" of a kinetic energy dominated "pre inflationary" universe. i.e. in terms of an inflaton $\dot{\phi}^2 \gg \left(P.E \sim V\right)$ [7] . The key finding which is in [11] is, that, if the kinetic energy is dominated by the "inflaton" that

$$K.E. \sim \left(E-V\right) \sim \dot{\phi}^2 \propto a^{-6}$$

(20)

This is done with the proviso that $w < -1$, in effect, what we are saying is that during the period of the "Planckian regime" we can seriously consider an initial density proportional to Kinetic energy, and call this K.E. as proportional to [7]

$$\rho_w \propto a^{-3(1-w)}$$

(21)

If we are where we are in a very small Planckian regime of space-time, we could, then say write Equation (21) as proportional to $g^* T^4$ [7] , with g^* initial degrees of freedom, and T the initial temperature as just before the onset of inflation. The question to ask, then is, what is the value of the initial degrees of freedom, and what is the temperature, T, at the start of expansion? For what it is worth, the starting supposition, is that there would then be a likelihood for an initial low temperature regime

MULTIVERSE, AND ANSWERING THE MUKHANOV HYPOTHESIS. INFLUENCE OF THE EINSTEIN SPACES

Here, the initial $a_0 \sim a_{initial} \sim 10^{-55}$, or so and so the density in Equation (21) at Planck time would, be proportional to the Planck Frequency [7]

$$\omega_P = \frac{1}{t_P} = \sqrt{\frac{c^5}{\hbar G}} = 1.85487 \times 10^{43} \text{ s}^{-1} \sim 1.85 \times 10^{43} \text{ Hz}$$

(22)

This is at the instant of Planck time. We can then ask what would be an initial time contribution before the onset of Planck time. i.e. does Equation (22) represent the initial value of graviton frequency?

This value of the frequency of a graviton, which would be red shifted enormously would be in tandem with an initial time step of as given by [12]

$$t_{initial} \approx \frac{1}{\sqrt{6\pi \rho_{initial}}} \sim \frac{a_0^2}{\sqrt{6\pi}}$$

(23)

This value for the initial time step would be probably lead to Pre Planckian time, i.e. smaller than 10^-43 seconds, which then leads us to consider, what would happen if a multi verse contributed to initial space-time conditions as seen in Equation (11) above. If the cosmic fluid approximation as given by Equation (12) were legitimate, and one could also look at Equation (13), then

$$m_{graviton} \geq \frac{2\hbar^2}{(\delta g_{tt})^2 l_P^2} \cdot \frac{(E-V)}{\Delta T_{tt}^2} \Rightarrow (\delta g_{tt})^2 \geq \frac{2\hbar^2}{m_{graviton} \cdot l_P^2} \cdot \frac{(E-V)}{\Delta T_{tt}^2}$$

(24)

But, then if one is looking at a multiverse, we first will start at the Penrose hypothesis for a cyclic conformal universe, starting with [13]

$$\hat{g}^{uv} = \Omega_{uv} g^{uv}$$

$$\Omega_{uv} \text{ (new-universe)} = \left(\Omega_{uv}^{-1} \text{ old-universe} \right)$$

i.e. $\Omega_{uv} \to \Omega_{uv}^{-1} \text{ (inversion)}$

(25)

However, in the multiverse contribution to Equation (12) above, we would have, that

$$\Omega_{uv}^{-1} \text{ old-universe} \to \frac{1}{N} \sum_{j=1}^{N} \left[\Omega_{uv}^{-1} \text{ (inversion)} \right]_j$$

(26)

So, does something like this hold? In a general sense?

$$(\delta g_{uv})^2 \Big|_{\text{initial}} \hat{g}^{uv} = \Omega_{uv} g^{uv}$$

$$\rightarrow \frac{1}{N} \sum_{j=1}^{N} \left[\Omega_{uv}^{-1}(\text{inversion})\right]_j g^{uv} \sim 1/\left[\beta M_{\text{Planck}}^2\right] + \varepsilon^+$$

(27)

If the fluid approximation as given in Equation (12) and Equation (13) hold, then Equation (27) conceivably could be identifiable as linkable to.

$$\delta t \Delta E \geq \frac{h}{(\delta g_{tt})} \equiv \frac{N \cdot h}{\left(\left[\sum_j \left[\Omega_{tt}^{-1}\right]_j\right] \cdot \delta g_{tt}\right)}$$

$$\Leftrightarrow \delta t \geq \frac{1}{\Delta E} \cdot \frac{N \cdot h}{\left(\left[\sum_j \left[\Omega_{tt}^{-1}\right]_j\right] \cdot \delta g_{tt}\right)} \approx \frac{m_{\text{graviton}} l_P \cdot l_P}{2 \cdot (E - V)}$$

(28)

If we could write, say

$$\frac{1}{\Delta E} \cdot \frac{N \cdot h}{\left(\left[\sum_j \left[\Omega_{tt}^{-1}\right]_j\right] \cdot \delta g_{tt}\right)} = \frac{N \cdot h / \Delta E}{\left(\sum_j \left[\Omega_{tt}^{-1} \cdot (\delta g_{tt})\right]_j\right)} \sim \frac{m_{\text{graviton}} l_P \cdot l_P}{2 \cdot (E - V)}$$

$$\Leftrightarrow \Delta E \sim 2 \cdot (E - V)$$

$$\&\quad m_{\text{graviton}} l_P \cdot l_P \sim \frac{N \cdot h}{\left(\sum_j \left[\Omega_{tt}^{-1} \cdot (\delta g_{tt})\right]_j\right)}$$

(29)

Then, if each j is the jth contribution of N "multiverse" contributions to a new single universe being nucleated, one could say that there was, indeed, likely an "averaging" and that the causal barrier which Mukhanov spoke of, as to each δt, and actually to each graviton entering into the present universe, one could mathematically average out the results of a sum up of each of the contributions from each prior to a present universe, according to

$$\frac{N \cdot h}{\left(\sum_j \left[\Omega_{tt}^{-1} \cdot (\delta g_{tt})\right]_j\right)} \equiv \frac{h}{\frac{1}{N} \sum_j \left[\Omega_{tt}^{-1} \cdot (\delta g_{tt})\right]_j}$$

(30)

If Equation (30) held, then we could then write

$$\delta t \geq \frac{1}{\Delta E} \cdot \frac{h}{\left(\frac{1}{N} \sum_j \left[\Omega_{tt}^{-1} \cdot \delta g_{tt}\right]_j\right)} \approx \frac{m_{\text{graviton}} l_P \cdot l_P}{2 \cdot (E - V)}$$

(31)

Instead, we have, Equation (28), and that it is safe to say that for each collapsing universe which might contribute to a re cycled universe that the following inequality is significant.

$$\frac{1}{N}\cdot\sum_{j}\left[\Omega_{tt}^{-1}\cdot(\delta g_{tt})\right]_{j}\neq\frac{1}{N}\cdot\sum_{j}\left[\Omega_{tt}^{-1}\right]_{j}\cdot\delta g_{tt}$$

(32)

Hence, the absence of an averaging procedure, due to a multiverse, would then rule against a causal barrier, as was maintained by Mukhanov, in his discussion with the author, in Marcel Grossman 14, in Italy. Then the possible approximation says of

$$\left(T_{tt}\right)^{2}\sim\omega_{\text{graviton}}^{2}\propto\beta M_{\text{Planck}}^{2}\sim\left(t_{\text{initial}}\approx\frac{1}{\sqrt{6\pi\rho_{\text{initial}}}}\sim\frac{a_{0}^{2}}{\sqrt{6\pi}}\right)^{-2}$$

(33)

Would not hold, and that in itself may lead to a breakdown of the Causal barrier hypothesis of Mukhanov, which the author emphatically disagreed with.

CONCLUSION. CONSIDERING EQUATION (6) AND EQUATION (11) IN LIEU OF EINSTEIN SPACE, AND FURTHER RESEARCH QUESTIONS

A way of solidifying the approach given here, in terms of early universe GR theory is to refer to Einstein spaces, via [14] as well as to make certain of the Stress energy tensor [15] as we can write it as a modified Einstein field equation. With, then $\aleph$ as a constant.

$$R_{ij}=\aleph g_{ij}$$

(34)

Here, the term in the Left hand side of the metric tensor is a constant, so then if we write, with R also a constant [15]

$$T_{ij}=-\frac{2}{\sqrt{-g}}\frac{\delta S}{\delta g_{ij}}=-\frac{1}{8\pi}\cdot\left[\aleph-R+\Lambda\right]\cdot g_{ij}$$

(35)

The terms, if we use the fluid approximation given by Equation (12) as well as the metric given in Equation (9) will then tend to a constant energy term on the RHS of Equation (35) as well as restricting i, and j, to t and t.

So as to recover, via the Einstein spaces, the seemingly heuristic argument given above. Furthermore when we refer to the Kinetic energy space as an inflaton where we assume that the potential energy is proportional to V, so as to allow us to write $\dot{\phi}^{2}\gg(P.E\sim V)$ [7] , we can also then utilize the following operator equation for the generation of an "inflaton field" given by the following set of equations

$$\phi(t) = \cos\left(t\sqrt{K}\right)f + \frac{\sin\left(t\sqrt{K}\right)}{\sqrt{K}}g$$

$$f(x) = \phi(0, x)$$

$$g(x) = \frac{\partial \phi(0, x)}{\partial t}$$

$$-\frac{\partial^2 \phi}{\partial t^2} = K\phi$$

(36)

In the case of the general elliptic operator K if we are using the Fulling reference, [16] in the case of the above Roberson-Walker metric, with the results that the elliptic operator, in this case become,

$$K = -\nabla^2 + \left(m^2 + \xi R\right) = -\sum_{i,j} \frac{\partial_i \left(g^{ij}\sqrt{|\det g|}\partial_j\right)}{\sqrt{|\det g|}} + \left(m^2 + \xi R\right)$$

$$\xrightarrow[i,j \to t,t]{} -\frac{\partial^2}{\partial t^2} + \left(m^2 + \xi R\right)$$

(37)

Then, according to [16], if R above, in Equation (37) is initially a constant, we will see then, if m is the inflation mass, that

$$\phi(t) = \cos\left(t\sqrt{K}\right)f - \frac{\partial^2}{\partial t^2} \to \omega^2$$

$$\Leftrightarrow \phi(t) = \cos\left(t\sqrt{\omega^2 + \left(m^2 + \xi R\right)}\right)$$

(38)

Then c_1 as an unspecified, for now constant will lead to a first approximation of a Kinetic energy dominated initial configuration, with details to be gleaned from [16] - [18] to give more details to the following equation, R here is linked to curvature of space-time, and m is an inflaton mass, connected with the field $\phi(t) = \cos\left(t\sqrt{K}\right)f$ with the result that

$$\dot{\phi}^2(t) \approx \left[\omega^2 + \left(m^2 + \xi R\right)\right]\cdot c_1 \gg V(\phi)$$

(39)

If the frequency, of say, Gravitons is of the order of Planck frequency as in Equation (22), then this term, would likely dominate Equation (39). More of the details of this will be worked out, and also candidates for the $V(\phi)$ will be ascertained, most likely, we will be looking the Rindler Vacuum as specified in [19] as well as also details of what is relevant to maintain local covariance in the initial space-time fields as given in [20].

Why is a refinement of Equation (39) necessary?

The details of the elliptic operator K will be gleaned from [16] - [18] whereas the details of inflaton $\dot{\phi}^2 \gg (P.E \sim V)$ [7] are important to get a refinement on the lower mass of the graviton as given by the left hand side of Equation (24). We hope to do this in the coming year. The mass, m, in Equation (37) for the inflaton, not the Graviton, so as to have links to the beginning of the expansion of the universe. We look to what Corda did, in [21] for guidance as to picking values of m relevant to early universe conditions.

Finally, as far as Equation (39) is concerned, there is one serious linkage issue to classical and quantum mechanics, which should be the bridge between classical and quantum regimes, as far as space time applicability. Namely, from Wald (19), if we look at first of all arbitrary operators, A and B

$$\left(\Delta A\right)^2 \cdot \left(\Delta B\right)^2 \geq \left(\frac{1}{2i}\langle[A,B]\rangle\right)$$

(40)

As we can anticipate, the Pre Planckian regime may the place to use classical mechanics, and then to bridge that to the Planckian regime, which would be quantum mechanical. Taking [19] again, this would lead to a sympletic structure via the following modification of the Hamilton equations of motion, namely we will from (19) get the following re write,

$$\frac{dq_\mu}{dt} = \frac{\partial H}{\partial p_\mu}, \quad \frac{dp_\mu}{dt} = -\frac{\partial H}{\partial q_\mu}$$

$$H = H\left(q_1, \cdots, q_n; p_1, \cdots, p_n\right)$$

$$y = \left(q_1, \cdots, q_n; p_1, \cdots, p_n\right)$$

$$\Omega^{\mu\nu} = 1, \text{ if } \nu = \mu + n$$

$$\Omega^{\mu\nu} = 0, \text{ otherwise}$$

$$\frac{dy^\mu}{dt} = \sum_{\nu=1}^{n} \Omega^{\mu\nu} \frac{\partial H}{\partial y^\nu}$$

(41)

Then there exists a re formulation of the Poisson brackets, as seen by

$$\{f, g\} = \Omega^{\mu\nu} \nabla_\mu f \nabla_\nu g$$

(42)

So, then the following, for classical observables, f, and g, we could write, by [19]

$$^\wedge : \Theta \to \hat{\Theta}$$

$$\Theta = \text{classical-observable}$$

$$\hat{\Theta} = \text{quantum-observable}$$

$$\hbar = 1$$

$$\left[\hat{f}, \hat{g}\right] = i \cdot \left(\widehat{\{f, g\}}\right)$$

(43)

Then, we could write, say Equation (40) and Equation (43) as

$$\left[\hat{f},\hat{g}\right]=i\cdot\widehat{\left(\{f,g\}\right)}$$

f = classical-observable

$\hat{f}$ = quantum-observable

$$\left(\Delta\hat{f}\right)^{2}\cdot\left(\Delta\hat{g}\right)^{2}\geq\left(\frac{1}{2i}\left(\left\langle\left[\hat{f},\hat{g}\right]\right\rangle\right)\right)=\left(\frac{1}{2}\left(\widehat{\left(\{f,g\}\right)}\right)\right)$$

(44)

If so, then we can set, in the interconnection between the Planck regime, and just before the Planck regime, say, by setting classical variables, as given by

$$f=-\frac{\left[\aleph-R+\Lambda\right]\cdot g_{tt}}{8\pi}$$

$$g=\delta g_{tt}$$

(45)

Then by utilization of Equation (44) we may be able to effect more precision in our early universe derivation, especially making use of derivational work, in addition as to what is given here, as to understand how to construct a very early universe partition function Z based upon the inter relationship between Equation (44) and Equation (45) so as to write up an entropy based upon, as given in [19]

$$S\left(\text{entropy}\right)=\ln Z+\beta E$$

(46)

If this program were affected, with a first principle construction of a partition function, we may be able to answer if Entropy were zero in the Planck regime, or something else, which would give us more motivation to examine the sort of partition functions as stated in [22] [23] . See Appendix A as to possible scenarios. Here keep in mind that in the Planck regime we have nonstandard physics. Appendix A indicates that due to the variation we have worked out in the Planckian regime of space-time that the initial entropy is not zero. The consequences of this show up in this paper's Appendix B, as to a specific formulation of the Ricci scalar. The consequences of Appendix A and Appendix B may be for a small cosmological constant, and large " Hubble expansion" that there would be an initially large magnitude of cosmological pressure, even if negative, which would give credence to a non-zero cosmological entropy, that if large negative pressure, even in the Pre Planckian regime will lead to a large ΔT_{tt} terms which would show up in Equation (1A), even if we used a partition function based upon Lattice Hamiltonians, as on page 135 of [26] which would usually in a lattice gauge arrangement would have considerably smaller contributions than ΔT_{tt}. Note the conditions of flat space, are that Equation (B9) almost vanishes due to the behavior of the numerator, no matter how small $a_{initial}^{2}$ is. The supposition is that the numerator becomes far smaller than $a_{initial}^{2}$ The initiation of conditions of flat space, is also the regime in which we think that non zero entropy is started, and Appendix C gives an initial estimate of what we think Entropy would be in the aftermath of the uncertainty relationship we have outlined in this article. i.e. to first order, $S_{\text{initial(graviton)}} \sim 10^{37}$. We finalize our treatment as of space-time fluctuations and geometry by considering the applications of Appendix D to graviton mass, and Appendix E to the Riemann-Penrose inequality for conditions as to a minimum frequency, as a consequence of cosmological evolution, and what it portrays as consequences for Electromagnetic fields. Appendix D and E give varying initial graviton masses as a starting point, with Appendix D giving a higher initial graviton mass than what is assumed as of today. Finally, Appendix F states a pre Planckian kinetic energy so the inflaton $\dot{\phi}^2 \gg (P.E \sim V)$ [7] . This last step, so important to our development will be considerably refined in future document.

We start the process of understanding the consequences of choosing the inflaton $\dot{\phi}^2 \gg (P.E \sim V)$ [7] as given in part in Appendix G and Appendix H.

The consequences of the above mentioned appendix entries are, mainly that if we wish to avoid the problems given in Appendix G and Appendix H that we really need to keep in mind the following:

1) Our uncertainty principle is fundamentally different from the Black hole commensurate uncertainty principles cited in Appendix G. They do not take into consideration the possibility that there may be Pre Planckian time, which may immensely impact the fluctuations in the metric tensor.

2) As an exercise, Appendix G shows that a highly restricted parameter space is required if we insist upon making our Pre Planckian uncertainty principle commensurate with the possibility that our metric Heisenberg Uncertainty principle (HUP) is in fact, giving us the flat space result which was brought up by Mukhanov, in Marcel Grossman 14. But it is so restrictive that we doubt it is actually mathematically a useful development

3) Appendix H gives us Equation (H1) which is the Pre Planckian Inflaton, which is of foundational importance in determination of if we have general relativity or some other gravitational theory, i.e. the issue of if there is an additional polarization. But to do that, we have to for reasons given in Appendix G, choose our parameter space, wisely. It is still not clear if there is a connection between Black hole physics, and avoiding the catastrophe of Bicep 2. For that much additional experimental work has to be done.

ACKNOWLEDGEMENTS

This work is supported in part by National Nature Science Foundation of China grant No. 11375279.

APPENDIX A: SCENARIOS AS TO THE VALUE OF ENTROPY IN THE BEGINNING OF SPACE-TIME NUCLEATION

We will be looking at inputs from page 290 of [23] so that if $E \sim M \sim \Delta T_{tt} \cdot \delta t_{\text{time}} \cdot \Delta A \cdot l_P$

$$S(\text{entropy}) = \ln Z + \frac{\left(E \sim \Delta T_{tt} \cdot \delta t \cdot \Delta A \cdot l_P\right)}{k_B T_{\text{temperature}}}$$

(1A)

And using Ng's infinite quantum statistics, we have to first approximation [24] [25]

$$S(\text{entropy}) \sim \ln Z + \frac{\left(\left(E \sim \Delta T_{tt}\right) \cdot \delta t \cdot \Delta A \cdot l_P\right)}{k_B T_{\text{temperature}}} \sim \ln Z + \left(\frac{h}{k_B T_{\text{temperature}} \delta g_{tt}}\right)$$

$$\xrightarrow[T_{\text{temperature}} \to \# \text{anything}]{} \left[S(\text{entropy}) \sim n_{\text{count}} \right] \neq 0$$

(2A)

This is due to a very small but non vanishing δg_{tt} with the partition functions covered by [23] , and also due to [24] [25] with n_{count} a non-zero number of initial "particle" or information states, about the Planck regime of space-time, so that the initial entropy is non zero.

APPENDIX B: CALCULATION OF THE RICCI TENSOR FOR A ROBERSON-WALKER SPACE-TIME, WITH ITS EFFECT UPON THE MEASUREMENT OF IF OR NOT A SPACE TIME, IS OPEN, CLOSED OR FLAT

We begin with Kolb and Turner [7] discussion of the Roberson-Walker metric, say page 49 with, if R is the Ricci scalar, and k the measurement of if we have a close, open, or flat universe, that if

$$a = a_{\text{initial}} \cdot \exp\left(H \cdot t\right)$$

(B1)

Then by [7]

$$H^2 = -\frac{k}{a^2} + \frac{8\pi G \rho}{3}$$

(B2)

$$3H^2 + \left[\frac{2k}{a^2} + \frac{R}{6}\right] = 0$$

(B3)

Leading to

$$a^2 = \frac{1}{k} \cdot \left[\frac{R}{6} + 8\pi G \rho \right] \tag{B4}$$

If $\rho = -p$ [7], then with a bit of algebra

$$|p| = \frac{1}{8\pi G} \cdot \left[\frac{R}{6} + (a_{\text{initial}})^2 \cdot \exp\left[\sqrt{\frac{4\Lambda}{3}} \cdot t_{\text{time}} \right] \right] \tag{B5}$$

Next, using [27], on page 47, at the boundary between Pre Planckian to Planckian space-time we will find

$$R = 8\pi \cdot \left(T_0^0 + T_1^1 + T_2^2 + T_3^3 \right) + 4\Lambda \xrightarrow[\text{Pre-Planckian-Conditions}]{} 8\pi \cdot \left(T_0^0 \right) + 4\Lambda \tag{B6}$$

Then, we can obtain

Right at the start of the Planckian era,

$$|p|_{\text{Planckian}} \sim \frac{1}{8\pi G} \cdot \left[\frac{8\pi \cdot \left(T_0^0 + T_1^1 + T_2^2 + T_3^3 \right) + 4\Lambda}{6} \right] \tag{B7}$$

The consequences of this would be that right after the entry into Planckian space time, that there would be the following change of pressure

$$|p|_{\text{Pre-Planckian}} = \frac{1}{8\pi G} \cdot \left[\frac{8\pi \cdot \left(T_0^0 \right) + 4\Lambda}{6} + (a_{\text{initial}})^2 \cdot \exp\left[\sqrt{\frac{4\Lambda}{3}} \cdot t_{\text{time}} \right] \right]$$

$$\Rightarrow |p|_{\text{Pre-Planckian}} \sim \frac{1}{8\pi G} \cdot \left| \frac{8\pi \cdot \left(T_0^0 \right) + 4\Lambda}{6} + 0^+ \right|$$

$$|p|_{\text{Planckian}} \sim \frac{1}{8\pi G} \cdot \left| \frac{8\pi \cdot \left(T_0^0 + T_1^1 + T_2^2 + T_3^3 \right) + 4\Lambda}{6} \right|$$

$$\Delta P = |p|_{\text{Planckian}} - |p|_{\text{Pre-Planckian}} \sim \left[\frac{\left(T_1^1 + T_2^2 + T_3^3 \right)}{6G} \right] \tag{B8}$$

Then, the change in the k term would be like, say, from Pre Planckian to Planckian space time

$$\Delta k = \frac{1}{a_{\text{initial}}^2} \cdot \left[8\pi G \left(\rho - \Delta P \right) \right] \tag{B9}$$

This goes almost to zero if the numerator shrinks far more than the denominator, even if the initial scale factor is of the order of 10^{-55} or so.

APPENDIX C: INITIAL ENTROPY, FROM FIRST PRINCIPLES

We are making use of the Padmanabhan publication of [28] [29] where we will make use of

$$\rho_\Lambda \approx \frac{G E_{system}^6}{c^8 h^4} \Leftrightarrow \Lambda \approx \frac{1}{l_{Planck}^2} \cdot \left(E_{system} / E_{Planck} \right)^6 \tag{C1}$$

Then, if E_{system} is for the energy of the Universe after the initiation of Equation (11) as a bridge between Pre Planckian, to Planckian physics regimes we could write, then

$$E_{system} \propto n_{gravitons} \cdot m_{graviton}$$

$$\Lambda \approx \frac{1}{l_{Radius-Universe-today}^2}$$

$$\Leftrightarrow m_{graviton} \sim 10^{-62} \text{ grams} \Rightarrow n_{gravitons} \sim 10^{37}$$

$$\Rightarrow S_{initial(graviton)} \sim 10^{37} \text{ at Planck time} \tag{C2}$$

The value of initial entropy, $S_{initial(graviton)} \sim 10^{37}$ should be contrasted with the entropy for the entire Universe as given in [30] below.

APPENDIX D: INFORMATION FLOW, GRAVITONS, AND ALSO UPPER BOUNDS TO GRAVITON MASS

Here we can view the possibility of considering the following, namely [31] is extended by [32] so we can we make the following identification?

$$N = N_{graviton} \big|_{r_H} = \frac{c^3}{G \cdot h} \cdot \frac{1}{\Lambda} \approx \frac{1}{\Lambda} \tag{D1}$$

Should the N above, be related to entropy, and Equation (8) this supposition has to be balanced against the following identification, namely, as given by T. Padmanabhan [28] [29]

$$\Lambda_{Einstein-Const.Padmanabhan} = 1/l_{Planck}^2 \cdot \left(E / E_{Planck} \right)^6 \tag{D2}$$

But should the energy in the numerator in Equation (D2) be given as say by (C2), of Appendix C, we have quintessence. then there would have been quintessence, i.e. variation in the "Einstein constant", which would have a large impact upon mass of the graviton, with a sharp decrease in g_* being consistent with an evolution to the ultra-light value of the Graviton, with initial frequencies of the order of say for wavelength values initially the size of an atom,

$$\omega_{\text{initial}}\big|_{r_H \sim \text{atomic-size}} \sim 10^{21} \text{ Hz}$$

(D3)

The final value of the frequency would be of a magnitude smaller than one Hertz, so as to have value of the mass of the graviton would be then of the order of 10^{-62} grams [10] , due to Equation (D2) approaching [31] below, namely

$$\Lambda_{\text{Einstein-Const.}} = 1/l^2_{\text{Radius-Universe}}$$

(D4)

Leading to the upper bound of the Graviton mass of about 10^{-62} grams [31] [32] in the present era

$$m_{\text{graviton}} = \frac{\hbar}{c} \cdot \sqrt{\frac{(2\Lambda)}{3}} \approx \sqrt{\frac{(2\Lambda)}{3}}$$

(D5)

Equation (D5) has a different value if the entropy/particle count is lower, as has been postulated in this note. But the value of Equation (D5) becomes the Graviton mass of about 10^{-62} grams [10] in the present era which is in line with the entropy being far larger in the present era [30]

APPENDIX E: APPLYING THE RIEMANNIAN PENROSE INEQUALITY WITH APPLICATIONS IN OUR FLUCTUATION

If from Giovannini [33] we can write

$$\delta g_{tt} \sim a^2(t) \cdot \phi \ll 1$$

(E1)

Refining the inputs from Equation (E1) means more study as to the possibility of a non-zero minimum scale factor [34] , as well as the nature of ϕ as specified by Giovannini [33] . We hope that this can be done as to give quantifiable estimates and may link the non-zero initial entropy to either Loop quantum gravity "quantum bounce" considerations [35] and/or other models which may presage modification of the sort of initial singularities of the sort given in [1] . Furthermore if the non-zero scale factor is correct, it may give us opportunities as to fine tune the parameters given in [34] below.

$$\alpha_0 = \sqrt{\frac{4\pi G}{3\mu_0 c}} B_0$$

$$\hat{\lambda} \left(\text{defined}\right) = \Lambda c^2 / 3$$

$$a_{min} = a_0 \cdot \left[\frac{\alpha_0}{2\hat{\lambda}\left(\text{defined}\right)} \left(\sqrt{\alpha_0^2 + 32\hat{\lambda}\left(\text{defined}\right) \cdot \mu_0 \omega \cdot B_0^2} - \alpha_0 \right) \right]^{1/4}$$

(E2)

where the following is possibly linkable to minimum frequencies linked to E and M fields [34], and possibly relic Gravitons

$$B > \frac{1}{2 \cdot \sqrt{10\mu_0 \cdot \omega}}$$

(E3)

So, now we investigate the question of applicability of the Riemann Penrose inequality which is [36], p431, which is stated as

Riemann Penrose Inequality: Let (M, g) be a complete, asymptotically flat 3-manifold with Non negative-scalar curvature, and total mass m, whose outermost horizon Σ has total surface area A. Then

$$m_{\text{total-mass}} \geq \sqrt{\frac{A_{\text{surface-Area}}}{16\pi}}$$

(E4)

And the equality holds, if (M, g) is isometric to the spatial isometric spatial Schwartzshield manifold M of mass m outside their respective horizons.

Assume that the frequency, say using the frequency of Equation (E3), and $A \approx A_{min}$ of Equation (E4) is employed. So then say we have, if we use dimensional analysis appropriately, that

$$\left(v = \text{velocity} \equiv c \right) = f\left(\text{frequency}\right) \times \lambda\left(\text{wavelength}\right)$$

$$\Rightarrow \omega \approx \omega_{\text{initial}} \sim \frac{c}{d_{min}} \sim \frac{1}{d_{min}} \bigg|_{c=1} \quad \& \quad d_{min} \sim A^{1/3} \propto a_{min}$$

(E5)

Assume that we also set the input frequency as to Equation (E3) as according to $10 < \zeta \leq 37$ i.e. does

$$\left(m_{\text{total-mass}} \sim 10^\zeta \cdot m_{\text{graviton}} \right)^2 \propto a_{min}^3 / 16\pi$$

$$\Leftrightarrow \omega \approx \omega_{\text{initial}} \sim \frac{1}{d_{min}} \sim \left(16\pi \times 10^\zeta \cdot m_{\text{graviton}} \right)^{-2/3}$$

(E6)

Our supposition is that Equation (E6) should give the same frequency as of Equation (D3) above. So if we have in

In doing this, this is a frequency input into Equation (E3) above where we are safely assuming a graviton mass of about [10]

$$m_{\text{total-mass}} \sim 10^{37} \cdot m_{\text{graviton}}$$

$$m_{\text{graviton}} \sim 10^{-62} \text{grams}$$

(E7)

Does the following make sense? i.e. look at, when $10 < \zeta \le 37$

$$\left(m_{\text{total-mass}} \sim 10^{\zeta} \cdot m_{\text{graviton}} \right)^2 \propto a_{\min}^3 / 16\pi$$

$$\Leftrightarrow \omega \approx \omega_{\text{initial}} \sim \frac{1}{d_{\min}} \sim \left(16\pi \times 10^{\zeta} \cdot m_{\text{graviton}} \right)^{-2/3}$$

(E8)

We claim that if this is an initial frequency and that it is connected with relic graviton production, that the minimum frequency would be relevant to Equation (E3), and may play a part as to admissible B fields

Note, if Appendix D is used, this makes a re do of Equation (E8) which is a way of saying that the graviton mass given by [10] no longer holds.

In either case, Equation (E8) and Equation (E3) in some configuration may argue for implementation of work the author did in reference [37] as to relic cylindrical GW, i.e. their allowed frequency and magnitude, so considered.

APPENDIX F: FIRST PRINCIPLE TREATMENT OF PRE PLANCKIAN KINETIC ENERGY SO THE INFLATON
$\dot{\phi}^2 \gg (P.E \sim V)$ [7]

We give this as a plausibility argument which undoubtedly will be considerably refined, but its importance cannot be overstated. i.e. this is for Pre inflationary, Pre Planckian physics, so as to get a lower bound to the Graviton mass. To do this, we look at what [7] is saying and also we will be enlisting a new reference, [38] , by Bojowald, and also Padmanbhan [39] as to details to put in, so as to confirm a dominance of Kinetic energy. Start with a Friedman equation of

$$\left(\frac{\dot{a}}{a} \right)^2 + \frac{k_{\text{curvature}}}{a^2} = \frac{4\pi G}{3} \cdot \frac{p_\phi^2}{a^6} + \Lambda$$

(F1)

We will treat, then the Hubble parameter, as

$$\left(\frac{\dot{a}}{a} \right) = H_{\text{initial}} \equiv \frac{2}{t \cdot \left(1 + \dfrac{P}{\rho} \right)} \xrightarrow{P = -\rho + \varepsilon^+} \frac{2}{t \cdot \left(\dfrac{\varepsilon^+}{\rho} \right)} \xrightarrow{t \to t_P} \frac{2\rho}{t_p \cdot \varepsilon^+}$$

(F2)

Now from Padmanabhan, [39] , we can write density, in terms of flux according to

$$\frac{d\rho}{dt} = \frac{1}{V^{(3)} = \text{Volume}} \cdot \left(A = \text{Area} \right) \cdot \left(\Im = \text{Flux} \right) \sim \frac{\left(\Im = \text{Flux} \right)}{l_P}$$

(F3)

Then using 463 of [39] , if T is temperature, here, then if N is the particle count in the flux region per unit time (say Planck time), as well as using the "ideal gas law" approximation, for superhot conditions

$$\frac{d\rho}{dt} = \frac{1}{V^{(3)} = \text{Volume}} \cdot (A = \text{Area}) \cdot (\Im = \text{Flux}) \sim \frac{(\Im = \text{Flux})}{l_p}$$

$$\rho \sim \frac{(\Im = \text{Flux})}{c}$$

$$\Rightarrow H = \frac{N}{\varepsilon^+} \cdot \frac{1}{V^{(4)} = \text{4-Dim Volume}} \cdot \sqrt{\frac{8}{\pi}} \sqrt{\frac{k_B T}{m_{\text{flux-particle}}}}$$
(F4)

Next, according to [38] , we can make the following substitution.

$$p_\phi = a^3 \cdot \dot{\phi}$$
(F5)

Therefore, if

$$\approx a^{-6} \cdot (12\pi G) \cdot V^{(4)} \left(\left[\frac{N}{\varepsilon^+} \cdot \frac{1}{V^{(4)} = \text{4-Dim Volume}} \cdot \sqrt{\frac{8}{\pi}} \sqrt{\frac{k_B T}{m_{\text{flux-particle}}}} \right]^2 + |\Lambda| \right)$$
(F6)

If the scale factor is very small, say of the order of $a = a_{\text{initial}} \sim 10^{-55}$, then no matter how fall the initial volume is, in four space (it cancels out in the first part of the brackets), it's easy to see then that $\dot{\phi}^2 \gg (P.E \sim V)$ [7] .

We will in the future add more structure to this calculation so as to confirm via a precise calculation that the lower bound to the graviton mass, is about 10^{-70} grams. This value of 10^{-70} grams is an approximation, via dimensional analysis and will be improved, by more exact calculations.

APPENDIX G: THE GENERALIZED UNCERTAINTY PRINCIPLE IN QUANTUM GRAVITY COMPARED WITH OUR HEISENBERG UNCERTAINTY PRINCIPLE FOR A METRIC IN PRE PLANCKIAN SPACE-TIME

We are looking here at what was done in [40] [41] and noting that in particular that the [40] calculation of fluctuations in energy as given by bounds given by Black hole physics, such that, if we pick Planck's constant $\hbar = 1$

$$\Delta E \geq \frac{1}{2 \cdot (\Delta x)} - \alpha \cdot \left(\frac{1}{2 \cdot (\Delta x)} \right)^2 + \alpha^2 \cdot 5 \cdot \left(\frac{1}{2 \cdot (\Delta x)} \right)^3$$
(G1)

Compare that with our given value of

$$\delta t \Delta E \geq \frac{\hbar}{\delta g_{tt}} \neq \frac{\hbar}{2}$$

Unless $\delta g_{tt} \sim O(1)$

(G2)

This should be compared with our value of equivalence between these two equations which demands

$$\left(\delta g_{tt}\right)^{-1'} \delta t^{-1} \doteq \frac{1}{2 \cdot (\Delta x)} - \alpha \cdot \left(\frac{1}{2 \cdot (\Delta x)}\right)^2 + \alpha^2 \cdot 5 \cdot \left(\frac{1}{2 \cdot (\Delta x)}\right)^3$$

$$\Rightarrow \delta g_{tt} \doteq \delta t^{-1} \cdot \frac{1}{\dfrac{1}{2 \cdot (\Delta x)} - \alpha \cdot \left(\dfrac{1}{2 \cdot (\Delta x)}\right)^2 + \alpha^2 \cdot 5 \cdot \left(\dfrac{1}{2 \cdot (\Delta x)}\right)^3}$$

(G3)

The collapse to a situation with ourselves recovering the standard Heisenberg Uncertainty relationship for fluctuations of energy is seen in, if Equation (G1) and Equation (G2) are both correct having then that

$$\delta t \Delta E \geq \frac{\hbar}{2} \doteq \frac{1}{2}$$

$$\Leftrightarrow \delta t^{-1} \sim \frac{1}{2 \cdot (\Delta x)} - \alpha \cdot \left(\frac{1}{2 \cdot (\Delta x)}\right)^2 + \alpha^2 \cdot 5 \cdot \left(\frac{1}{2 \cdot (\Delta x)}\right)^3$$

(G3)

Here, we want the situation for which we would have any time situation with the fluctuation of time, going to a very small number, and that the inverse fluctuation in time going to infinity would be, trivially due to, if Δx is of Planck length, obtaining α for which.

$$\alpha \approx \frac{2(\Delta x)}{5}$$

(G4)

It's an equation for α, with a vanishingly small contribution for α. i.e. we would have, to first order $\alpha \sim \varepsilon^+$, i.e. α being very small. But that in turn would require, to first order

$$\delta t^{-1} \sim \frac{1}{2 \cdot (\Delta x)}$$

(G5)

This would be equivalent to, then setting

$$\delta g_{tt} \sim O(1)$$

(G6)

Then by necessity, we would want to have a situation for which to have a more general situation as given in our document for a

$$\alpha \neq \frac{2(\Delta x)}{5} \tag{G7}$$

In fact, to reconcile Equation (G1) and Equation (G2) in the case of recovering a

$$\delta g_{tt} \sim \text{small-value} \neq O(1) \tag{G8}$$

That not only would α obey Equation (G7) that it would likely be fairly large.

The situation as given by L. Crowell in [41] as it is attuned to dimensional analysis, as given in

$$\Delta x = \sqrt{2\pi R c T_P} \tag{G9}$$

Here, R is the radius of a sphere for the origins of an emitted wave, which is in turn requiring R to be extraordinarily small. i.e. we recover the inputs for our analysis of [40] as it applies to our document but only if we have extremely sharp restraints upon R, if we wish to have fidelity with Equation (G4) and Equation (G5) in the sense of recovery of the traditional Heisenberg relations. T_P is a Planck time interval as given in [41] It is extremely small, commensurate with Equation (G9) being approximately Planck Length in value.

The problem with Equation (G9) is that there is no provision given as to Pre Planckian length values, and that it is restricted, dimensionally to Planckian Length and temperature, with no clue given as to what happens before a Planck length.

APPENDIX H: CONSIDERATIONS AS TO BICEP 2, THE MATTER OF SCALAR-TENSOR POLARIZATIONS AS AN ALTERNATIVE TO GENERAL RELATIVITY AND ALTERNATE GRAVITATIONAL THEORIES. AND EXPERIMENTAL TESTS OF GENERAL RELATIVITY VIA INTERFEROMETRIC METHODS

Quoting from the Authors' recent publication [42] .

From [43] we have the following to consider, namely trying to determine restraints upon the nature of gravity, i.e. is it consistent with General relativity or do we have an alternative situation as given in the following quote. We hope that getting a consistent model of inflaton physics will help clarify the following alternatives

Quote, in [42] of the result given in [43] :

This fact rules out the possibility of treating gravitation like other quantum theories, and precludes the unification of gravity with other interactions. At the present time, it is not possible to realize a consistent Quantum Gravity Theory which leads to the unification of gravitation with the other forces [17] [18] . On the other hand, one can define Extended Theories of Gravity those semi classical theories where the Lagrangian is modified, in respect to the standard Einstein-Hilbert gravitational Lagrangian, adding high-order terms in the curvature invariants (terms like R2, etc…) or terms with scalar fields non minimally coupled to geometry (terms like φ2R) [17] [18] .

End of quote from [43] .

We then will cite what is in [42] i.e. namely that our uncertainty relationship leads to inflaton physics, as given in the following quote.

Quote, from [42]

Needless to say we will require careful analysis of the result as given in reference [42] that

$$\delta t \Delta E = \frac{\hbar}{\delta g_{tt}} \equiv \frac{\hbar}{a^2(t) \cdot \phi}$$

$$\Leftrightarrow \phi(\text{initial}) \approx \frac{1}{t(\text{Planck})} \cdot \frac{1}{\left[a(\text{initial-start}) \sim 10^{-55} \right]^2 \cdot \left[\omega(\text{initial}) \approx 10^{30} (\text{or-more}) \text{Hertz} \right]} \quad \text{(H1)}$$

This enormous value for the inflaton, initially, needs to be examined further. It further should be linked to Corda's pioneering work with "gravity's breath", i.e. traces of the inflaton as given by [21] [44] and is the justification of Equation (H1) above. We can use this to determine what to make of the stochastic background of pre space time physics.

Next, Avoiding the Bicep 2 mistake. What we can do with Equation (H1)

Following [42] [43] what we are doing is examining the stochastic regime of space-time where the following holds.

Omni-directional gravitational wave background radiation could arise from fundamental processes in the early Universe, or from the superposition of a large number of signals with a point-like origin. Examples of the former include parametric amplification of gravitational vacuum fluctuations during the inflationary era, termination of inflation through axion decay or resonant preheating, Pre-Big Bang models inspired by string theory, and phase transitions in the early Universe; the observation of a primordial background would give access to energy scales of 10 to the 9 power, up to 10 to the 10 power GeV, well beyond the reach of particle accelerators on Earth

Needless to say though, we need above all to avoid getting many multiple stochastic signals, in what we process for primordial gravitational waves, and to use, instead tests to avoid getting dust signals which is what doomed Bicep 2, i.e. as was made very clear in [42] [45] [46] .

i.e. the problem is in avoiding multiple stochastic signals, and this is explained in the conclusion of [42] . But to obtain what is in [42] , Equation (H1) has to be thoroughly understood, and Equation (H1) is commensurate with the details as cited in Equation (G3) to Equation (G7) which have to be vetted experimentally. i.e. the uncertainty principle as cited in Equation (H1) leads to an inflaton which will allow us to determine if a third Polarization exists, as in scalar-tensor gravity, or the more traditional considerations given in [42] [43] .

This in turn may allow understanding if our document is commensurate with the considerations given in [47] .

REFERENCES

1. Will, C. (2015) Was Einstein Right? A Centenary Assessment. In: Ashtekar, A., Berger, B., Isenberg, J. and MacCallum, M., Eds., General Relativity and Gravitation: A Centennial Perspective, Cambridge University Press, Cambridge, 49-96.http://dx.doi.org/10.1017/cbo9781139583961.004

2. Will, C. (2014) The Confrontation between General Relativity and Experiment.http://relativity.livingreviews.org/Articles/lrr-2014-4/download/lrr-2014-4Color.pdf

3. Downes, T.G. and Milburn, G.J. (2012) Optimal Quantum Estimation for Gravitation.http://arxiv.org/abs/1108.5220

4. Unruh, W.G. (1986) Why Study Quantum Theory? Canadian Journal of Physics, 64, 128-130. http://dx.doi.org/10.1139/p86-019

5. Unruh, W.G. (1986) Erratum: Why Study Quantum Gravity? Canadian Journal of Physics, 64, 1453. http://dx.doi.org/10.1139/p86-257

6. Giovannini, M. (2008) A Primer on the Physics of the Cosmic Microwave Background. World Press Scientific, Hackensack. http://dx.doi.org/10.1142/6730

7. Kolb, E.W. and Turner, M.S. (1990) The Early Universe. Addison-Wesley Publishing Company, The Advanced Book Program, Redwood City.

8. Barbour, J. (2009) The Nature of Time. http://arxiv.org/pdf/0903.3489.pdf

9. Barbour, J. (2010) Shape Dynamics: An Introduction. In: Finster, F., Muller, O., Nardmann, M., Tolksdorf, J. and Zeidler, E., Eds., Quantum

Field Theory and Gravity, Conceptual and Mathematical Advances in the Search for a Unified Framework, Birkhauser, Springer-Verlag, London, 257-297.

10.   Goldhaber, A. and Nieto, M. (2010) Photon and Graviton Mass Limits. Reviews of Modern Physics, 82, 939-979. http://arxiv.org/abs/0809.1003http://dx.doi.org/10.1103/RevModPhys.82.939

11.   Handley, W.J., Brechet, S.D., Lasenby, A.N. and Hobson, M.P. (2014) Kinetic Initial Conditions for Inflation. http://arxiv.org/pdf/1401.2253v2.pdf

12.   Walecka, J.D. (2008) Introduction to Modern Physics, Theoretical Foundations. World Press Scientific Co, Pte. Ltd., Singapore.

13.   Penrose, R. (2010) Cycles of Time: An Extraordinary New View of the Universe. The Bodley Head, London.

14.   Petrov, A.Z. (1969) Einstein Spaces. Pergamum Press, Oxford and London.http://dx.doi.org/10.1016/b978-0-08-012315-8.50007-0

15.   Gorbunov, D. and Rubakov, V. (2011) Introduction to the Theory of the Early Universe, Cosmological Perturbations and Inflationary Theory. World Scientific Publishing Pte. Ltd, Singapore.

16.   Fulling, S.A. (1991) Aspects of Quantum Field Theory in Curved Spacetime (London Mathematical Society Student Texts). Cambridge University Press, Cambridge.

17.   Gutfreund, H. and Renn, J. (2015) The Road to Relativity, the History and Meaning of Einstein's "The Foundation of General Relativity", (Featuring the Original Manuscript of Einstein's Masterpiece). Princeton University Press, Princeton and Oxford.http://dx.doi.org/10.1515/9781400865765

18.   Griffiths, J. and Podolsky, J. (2009) Exact Space-Times in Einstein's General Relativity. Cambridge Monographs on Mathematical Physics, Cambridge.http://dx.doi.org/10.1017/CBO9780511635397

19.   Wald, R.M. (1994) Quantum Field Theory in Curved Space-Time and Black Hole Thermodynamics. University of Chicago Press, Chicago.

20.   Fredenhagen, K. and Rejzner, K. (2010) Local Covariance and Background Independence. In: Finster, F., Muller, O., Nardmann, M., Tolksdorf, J. and Zeidler, E., Eds., Quantum Field Theory and Gravity, Conceptual and Mathematical Advances in the Search for a Unified Framework, Birkhauser, Springer-Verlag, London, 15-23.

21.   Corda, C. (2012) Primordial Gravity's Breath. Electronic Journal of Theoretical Physics, 9, 1-10. http://arxiv.org/abs/1110.1772

22. Gilen, S. and Oriti, D. (2010) Discrete and Continuum Third Quantization of Gravity. In: Finster, F., Muller, O., Nardmann, M., Tolksdorf, J. and Zeidler, E., Eds., Quantum Field Theory and Gravity, Conceptual and Mathematical Advances in the Search for a Unified Framework, Birkhauser, Springer-Verlag, London, 41-64.

23. Birrell, N.D. and Davies, P.C.W. (1982) Quantum Fields in Curved Space. Cambridge Monographs on Mathematical Physics, Cambridge University Press, London.http://dx.doi.org/10.1017/CBO9780511622632

24. Jack Ng, Y. and Jack, Y. (2007) Holographic Foam, Dark Energy and Infinite Statistics. Physics Letters B, 657, 10-14. http://dx.doi.org/10.1016/j.physletb.2007.09.052

25. Jack Ng, Y. (2008) Space-Time Foam: From Entropy and Holography to Infinite Statistics and Nonlocality. Entropy, 10, 441-461. http://dx.doi.org/10.3390/e10040441

26. Hambler, H. (2009) Quantum Gravitation, the Feynman Path Integral Approach. Springer-Verlag, Berlin.

27. Ciufolini, I. and Wheeler, J. (1995) Gravitation and Inertia. Princeton Series in Physics, Princeton University Press, Princeton.

28. Padmanabhan, T. (2003) Cosmological Constant—The Weight of the Vacuum.http://arxiv.org/abs/hep-th/0212290

29. Padmanabhan, T.http://ned.ipac.caltech.edu/level5/Sept02/Padmanabhan/Pad1_2.html.

30. Egan, C. and Lineweaver, C.H. (2010) A Larger Estimate of the Entropy of the Universe. The Astrophysical Journal, 710, 1825-1834.http://www.mso.anu.edu.au/~charley/papers/EganLineweaverApJOnline.pdfhttp://dx.doi.org/10.1088/0004-637X/710/2/1825

31. Ali, A.F. and Das, S. (2015) Cosmology from Quantum Potential. Physics Letters B, 741, 276-279. http://dx.doi.org/10.1016/j.physletb.2014.12.057

32. Haranas, I. and Gkigkitzis, I. (2014) The Mass of Graviton and Its Relation to the Number of Information According to the Holographic Principle. International Scholarly Research Notices, 2014, Article ID: 718251.http://www.hindawi.com/journals/isrn/2014/718251/

33. Giovannini, M. (2008) A Primer on the Physics of the Cosmic Microwave Background. World Press Scientific, Hackensack. http://dx.doi.org/10.1142/6730

34. Camara, C.S., de Garcia Maia, M.R., Carvalho, J.C. and Lima, J.A.S. (2004) Nonsingular FRW Cosmology and Non Linear Dynamics. http://arxiv.org/abs/astro-ph/0402311

35. Rovelli, C. and Vidotto, F. (2015) Covariant Loop Quantum Gravity: An Elementary Introduction to Quantum Gravity and Spin foam Theory. Cambridge University Press, Cambridge.

36. Galloway, G., Miao, P. and Schoen, R. (2015) Initial Data and the Einstein Constraints. In: Ashtekar, A., Berger, B., Isenberg, J. and MacCallum, M., Eds., General Relativity and Gravitation: A Centennial Perspective, Cambridge University Press, Cambridge, 412-448.

37. Wen, H., Li, F.Y., Fang, Z.Y. and Beckwith, A. (2014) Impulsive Cylindrical Gravitational Wave: One Possible Radiative Form Emitted from Cosmic Strings and Corresponding Electromagnetic Response. http://arxiv.org/abs/1403.7277

38. Bojowald, M. (2012) A Momentous Arrow of Time. In: Mersini, L. and Vaas, R., Eds., The Arrows of Time: A Debate in Cosmology, Springer Verlag, Berlin, 169-189.http://arxiv.org/pdf/0910.3200.pdf

39. Padmanabhan, T. (2010) Gravitation, Foundations and Frontiers. Cambridge University Press, Cambridge. http://dx.doi.org/10.1017/CBO9780511807787

40. Ali, A.F., Khalil, M.M. and Vagenas, E.C. (2015) Minimal Length in Quantum Gravity and Gravitational Measurements. http://arxiv.org/abs/1510.06365

41. Crowell, L. (2015) Topology of States on a Black Hole Event Horizon. Electronic Journal of Theoretical Physics, 12, 211-218.

42. Beckwith, A. (2016) Gedanken Experiment (Thought Experiment) about Gravo-Electric and Gravo-Magnetic Fields, and the Link to Gravitons and Gravitational Waves in the Early Universe. Recently Accepted Article in JHEPGC.

43. Corda, C. (2009) Interferometric Detection of Gravitational Waves: The Definitive Test for General Relativity. International Journal of Modern Physics D, 18, 2275-2282.http://arxiv.org/abs/0905.2502 http://dx.doi.org/10.1142/S0218271809015904

44. Corda, C. (2007) A Longitudinal Component in Massive Gravitational Waves Arising from a Bimetric Theory of Gravity. Astroparticle Physics, 28, 247-250.http://arxiv.org/abs/0811.0985 http://dx.doi.org/10.1016/j.astropartphys.2007.05.009

45. Cowen, R. (2015) Gravitational Waves Discovery Now Officially Dead; Combined Data from South Pole Experiment BICEP2 and Planck Probe Point to Galactic Dust as Confounding Signal. http://www.nature.com/news/gravitational-waves-discovery-now-officially-dead-1.16830

46. Cowen, R. (2014) Full-Galaxy Dust Map Muddles Search for Gravitational Waves.http://www.nature.com/news/full-galaxy-dust-map-muddles-search-for-gravitational-waves-1.15975

47. Beckwith, A. (2016) Non Linear Electrodynamics Contributing to a Minimum Vacuum Energy ("Cosmological Constant") Allowed in Early Universe Cosmology. Journal of High Energy Physics, Gravitation and Cosmology, 2, 25-32.

Chapter 10

CLASSICAL AND RELATIVISTIC FLUX OF ENERGY CONSERVATION IN ASTROPHYSICAL JETS

Lorenzo Zaninetti

Physics Department, Turin, Italy

ABSTRACT

The conservation of the energy flux in turbulent jets which propagate in the intergalactic medium (IGM) allows deducing the law of motion in the classical and relativistic cases. Three types of IGM are considered: constant density, hyperbolic and inverse power law decrease of density. An analytical law for the evolution of the magnetic field along the radio-jets is deduced using a linear relation between the magnetic pressure and the rest density. Astrophysical applications are made to the centerline intensity of synchrotron emission in NGC315 and to the magnetic field of 3C273.

INTRODUCTION

The analysis of turbulent jets in the laboratory offers the possibility of applying the theory of turbulence to some well defined experiments, see [1] [2] . The experiments of Reynolds can be seen in [4] . Analytical results for the theory of turbulent jets can be found in [4] - [7] . Recently the analogy between laboratory jets and extragalactic radio-jets has been pointed out, see [8] [9] . We briefly recall that the theory of "round turbulent jets" can be defined in terms of the velocity at the nozzle, the diameter of the nozzle, and the viscosity, see Section 5 in [6] ; as an example the gradients in pressure are not considered. The application of the theory of turbulence to extragalactic radio-jets produces a great number of questions to be solved because we do not observe the turbulent phenomena but the radio features which have properties similar to the laboratory's turbulent jets, i.e. similar opening angles. We now pose the following questions.

- Is it possible to apply the conservation of the flux of energy in order to derive the equation of motion for radio-jets in the cases of constant and variable density of the surrounding medium?

- Can we extend the conservation of the flux of energy to the relativistic regime?
- Can we model the behaviour of the magnetic field and the intensity of synchrotron emission as functions of the distance from the parent nucleus?
- Can we model the back reaction on the equation of motion for turbulent jets due to radiative losses?

In order to answer these questions, we derive the differential equations which model the classical and relativistic conservation of the energy flux for a turbulent jet in the presence of different types of medium, see Sections 2 and 3. Section 4 presents classical and relativistic parametrizations of the radiative losses as well as the evolution of the magnetic field.

ENERGY CONSERVATION

The conservation of the energy flux in a turbulent jet requires the perpendicular section to the motion along the Cartesian x-axis, A

$$A(r) = \pi r^2 \tag{1}$$

where r is the radius of the jet. The section A at position x_0 is

$$A(x_0) = \pi \left(x_0 \tan\left(\frac{\alpha}{2}\right) \right)^2 \tag{2}$$

where α is the opening angle and x_0 is the initial position on the x-axis. At position x we have

$$A(x) = \pi \left(x \tan\left(\frac{\alpha}{2}\right) \right)^2. \tag{3}$$

The conservation of energy flux states that

$$\frac{1}{2}\rho(x_0)v_0^3 A(x_0) = \frac{1}{2}\rho(x)v(x)^3 A(x) \tag{4}$$

where $v(x)$ is the velocity at position x and $v_0(x_0)$ is the velocity at position x_0, see Formula A28 in [10] .

The selected physical units are pc for length and yr for time; with these units, the initial velocity v_0 is expressed in pc · yr^{-1}, 1 yr = 365.25 days. When the initial velocity is expressed in km · s^{-1}, the multiplicative factor 1.02×10^{-6} should be applied in order to have the velocity expressed in pc · yr^{-1}.

Constant Density

In the case of constant density of the intergalactic medium (IGM) along the x-direction, the law of conservation of the energy flux, as given by Equation (4), can be written as a differential equation

$$\left(\frac{d}{dt}x(t)\right)^3 \left(x(t)\right)^2 - v_0^3 x_0^2 = 0.$$

$$(5)$$

The analytical solution of the previous differential equation can be found by imposing $x = x_0$ at $t = 0$,

$$x(t) = \frac{1}{3} 3^{2/5} \sqrt[5]{x_0^2 (5tv_0 + 3x_0)^3}.$$

$$(6)$$

The asymptotic approximation is

$$x(t) \sim \frac{1}{3} 3^{2/5} 5^{3/5} \sqrt[5]{v_0^3 x_0^2} t^{3/5}.$$

$$(7)$$

The velocity is

$$v(t) = \frac{3^{2/5} x_0^2 (5tv_0 + 3x_0)^2 v_0}{\left(x_0^2 (5tv_0 + 3x_0)^3\right)^{4/5}}$$

$$(8)$$

and its asymptotic approximation

$$v(t) \sim \frac{1}{5} \frac{3^{2/5} \sqrt[5]{125} x_0^2 v_0^3 (t^{-1})^{2/5}}{\left(v_0^3 x_0^2\right)^{4/5}}.$$

$$(9)$$

The velocity as a function of the distance is

$$v(x) = \frac{x_0^{2/3} v_0}{x^{2/3}}.$$

$$(10)$$

A first comparison can be made with the laboratory data on turbulent jets of [11] where the velocity of the turbulent jet at the nozzle diameter, $D_j = 1$, is $v_0 = 2.53 \text{ m} \cdot \text{s}^{-1}$ and at $D_j = 50$ the centerline velocity is $v = 0.314 \text{ m} \cdot \text{s}^{-1}$. The formula (10) with $x_0 = 1$ and $x = 50$ gives an averaged velocity of $v = 0.186 \text{ m} \cdot \text{s}^{-1}$ which multiplied by 2 gives $v = 0.372 \text{ m} \cdot \text{s}^{-1}$. This multiplication by 2 has been done because the turbulent jet develops a profile of velocity in the direction perpendicular to the jet's main axis and therefore the centerline velocity is approximately double that of the averaged velocity. The transit time, t_{tr}, necessary to travel a distance of x_{max} can be derived from Equation (6)

$$t_{tr} = \frac{3\sqrt[3]{x_{max}^2 x_0 x_{max}} - 3x_0^2}{5x_0 v_0}.$$

(11)

An astrophysical test can be performed on a typical distance of 15 kpc relative to the jets in 3C 31, see Figure 2 in [12]. On inserting $x = 15,000$ pc $= 15$ kpc , $x_0 = 100$ pc, and $v_0 = 10,000$ km·s^{-1} we obtain a transit time of $t_{tr} = 2.488 \times 10^7$ yr .

The rate of mass flow at the point x, $\dot{m}(x)$, is

$$\dot{m}(x) = \rho v(x) \pi \left(x \tan\left(\frac{\alpha}{2}\right) \right)^2$$

(12)

and the astrophysical version is

$$\dot{m}(x) = 0.0237 n x^{4/3} \left(\tan(\alpha/2) \right)^2 x_0^{2/3} \beta_0 \frac{M_\odot}{yr}$$

(13)

where x and x_0 are expressed in pc, n is the number density of protons expressed in particles cm^{-3}, $M_\odot$ is the

solar mass and $\beta_0 = \frac{v_0}{c}$. The previous formula indicates that the rate of transfer of particles is not constant

along the jet but increases $\propto x^{4/3}$.

An Hyperbolic Profile of the Density

Now the density is assumed to decrease as

$$\rho = \rho_0 \left(\frac{x_0}{x} \right)$$

(14)

where $\rho_0 = 0$ is the density at $x = x_0$. The differential equation that models the energy flux is

$$x_0 x(t) \left(\frac{d}{dt} x(t) \right)^3 - v_0^3 x_0^2 = 0$$

(15)

and its analytical solution is

$$x(t) = \frac{1}{3} \sqrt[4]{3} \sqrt[4]{x_0 \left(4tv_0 + 3x_0 \right)^3}.$$

(16)

The asymptotic approximation is

$$x(t) \sim \frac{2}{3} \sqrt[4]{3} \sqrt{2} \sqrt[4]{v_0^3 x_0} t^{3/4}.$$

(17)

The analytical solution for the velocity is

$$v(t) = \frac{\sqrt[4]{3}x_0 \left(4tv_0 + 3x_0\right)^2 v_0}{\left(x_0 \left(4tv_0 + 3x_0\right)^3\right)^{3/4}}$$

(18)

and its asymptotic approximation is

$$v(t) \sim \frac{1}{4} \frac{\sqrt[4]{3}\sqrt[4]{64}x_0 v_0^3 \sqrt[4]{t}^{-1}}{\left(v_0^3 x_0\right)^{3/4}}.$$

(19)

The transit time can be derived from Equation (16)

$$t_{tr} = \frac{3\sqrt[3]{x_{max} x_0^2} x_{max} - 3x_0^2}{4x_0 v_0}$$

(20)

and with $x = 15,000 \text{ pc} = 15 \text{ kpc}$, $x_0 = 100 \text{ pc}$, and $v_0 = 10,000 \text{ km·s}^{-1}$ as in Section 2.1, we have $t_{tr} = 5.848 \times 10^6 \text{ yr}$.

An Inverse Power Law Profile of the Density

Here, the density is assumed to decrease as

$$\rho = \rho_0 \left(\frac{x_0}{x}\right)^\delta$$

(21)

where ρ_0 is the density at $x = x_0$. The differential equation which models the energy flux is

$$\frac{1}{2}\left(\frac{x_0}{x}\right)^\delta \left(\frac{d}{dt}x(t)\right)^2 x^2 - \frac{1}{2}v_0^2 x_0^2 = 0.$$

(22)

There is no analytical solution, and we simply express the velocity as a function of the position, x,

$$v(x) = \frac{x_0 v_0}{x} \frac{1}{\sqrt{\left(\frac{x_0}{x}\right)^\delta}}$$

(23)

see Figure 1

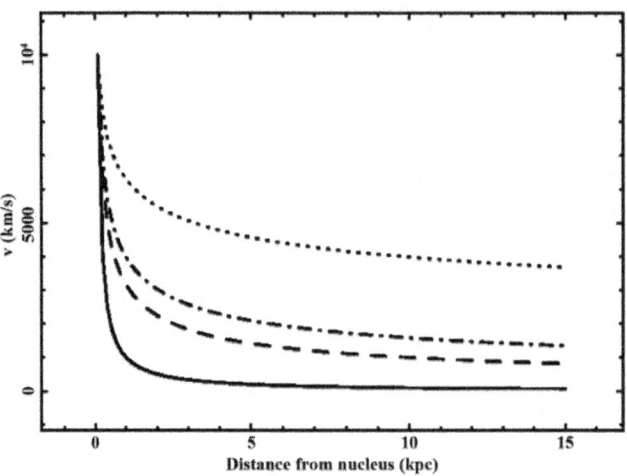

Figure 1. Classical velocity as a function of the distance from the nucleus when $x_0 = 100\,\text{pc}$ and $v_0 = 10{,}000\,\text{km}\cdot\text{s}^{-1}$: $\delta = 0$ (full line), $\delta = 1$ (dashes), $\delta = 1.2$ (dot-dash-dot-dash) and $\delta = 1.6$ (dotted).

The rate of mass flow at the point x is

$$\dot{m}(x) = \rho_0 \sqrt{\left(\frac{x_0}{x}\right)^{\delta}} \, \pi x \left(\tan\left(\alpha/2\right)\right)^2 x_0 v_0 \tag{24}$$

and the astrophysical version is

$$\dot{m}(x) = 0.0237 n_0 \sqrt{\left(1.0\frac{x_0}{x}\right)^{\delta}} \, x \left(\tan\left(\alpha/2\right)\right)^2 x_0 \beta_0 \frac{M_\odot}{yr} \tag{25}$$

where n_0 is the number density of protons expressed in particles cm^{-3} at x_0. The previous formula indicates

that the rate of transfer of particles scales $\propto x^{1-\frac{1}{2}\delta}$ and therefore at $\delta = 2$ is constant.

RELATIVISTIC TURBULENT JETS

The conservation of the energy flux in special relativity (SR) in the presence of a velocity v along one direction states that

$$A(x)\frac{1}{1-\frac{v^2}{c^2}}(e_0 + p_0)v = cost \tag{26}$$

where $A(x)$ is the considered area in the direction perpendicular to the motion, c is the speed of light, $e_0 = c^2 \rho$ is the energy density in the rest frame of the moving fluid, and p_0 is the pressure in the rest frame of the moving fluid, see formula A31 in [10]. In accordance with the current models of classical turbulent jets, we insert $p_0 = 0$ and the conservation law for relativistic energy flux is

$$\rho c^2 v \frac{1}{1 - \dfrac{v^2}{c^2}} A(x) = cost.$$

(27)

Our physical units are pc for length and yr for time, and in these units, the speed of light is $c = 0.306 \, \text{pc} \cdot \text{yr}^{-1}$. A discussion of the mass-energy equivalence principle in fluids can be found in [13].

Constant Density in SR

The conservation of the relativistic energy flux when the density is constant can be written as a differential equation

$$\rho c^2 \left(\frac{\mathrm{d}}{\mathrm{d}t} x(t) \right) \pi (x(t))^2 \left(\tan\left(\frac{\alpha}{2} \right) \right)^2 \left(1 - \frac{\left(\frac{\mathrm{d}}{\mathrm{d}t} x(t) \right)^2}{c^2} \right)^{-1}$$

$$- \rho c^2 v_0 \pi x_0^2 \left(\tan\left(\frac{\alpha}{2} \right) \right)^2 \left(1 - \frac{v_0^2}{c^2} \right)^{-1} = 0.$$

(28)

An analytical solution of the previous differential equation at the moment of writing does not exist but we can provide a power series solution of the form

$$x(t) = a_0 + a_1 t + a_2 t^2 + a_3 t^3 + \cdots$$

(29)

see [14] [15]. The coefficients a_n up to order 4 are

$a_0 = x_0$

$a_1 = v_0$

$$a_2 = \frac{1}{3} \frac{v_0^3 \left(5c^6 - 11c^4 v_0^2 + 3c^2 v_0^4 + 3v_0^6 \right)}{x_0^2 \left(c^2 + v_0^2 \right) \left(c^4 + 2c^2 v_0^2 + v_0^4 \right)}$$

$$a_3 = \frac{1}{3} \frac{v_0^3 \left(5c^6 - 11c^4 v_0^2 + 3c^2 v_0^4 + 3v_0^6 \right)}{x_0^2 \left(c^2 + v_0^2 \right) \left(c^4 + 2c^2 v_0^2 + v_0^4 \right)}.$$

(30)

In order to find a numerical solution of the above differential equation we isolate the velocity from Equation (28)

$$v(x;x_0,\beta_0,c) = \frac{1}{2}\frac{\left(\beta_0^2 x^2 - x^2 + \sqrt{x^4\beta_0^4 - 2x^4\beta_0^2 + 4\beta_0^2 x_0^4 + x^4}\right)c}{\beta_0 x_0^2}$$

(31)

where $\beta_0 = \frac{v_0}{c}$ and separate the variables

$$\int_{x_0}^x 2\frac{\beta_0 x_0^2}{\left(\beta_0^2 x^2 - x^2 + \sqrt{x^4\beta_0^4 - 2x^4\beta_0^2 + 4\beta_0^2 x_0^4 + x^4}\right)c}\,dx = \int_0^t dt.$$

(32)

The indefinite integral on the left side of the previous equation has an analytical expression

$$I(x;\beta_0,c,x_0) = \frac{AN}{AD}$$

(33)

where

$$AN = 2\beta_0^3 x_0^6 \sqrt{2}\sqrt{4 - 2\frac{i\beta_0 x^2}{x_0^2} + 2\frac{ix^2}{\beta_0 x_0^2}}\sqrt{4 + 2\frac{i\beta_0 x^2}{x_0^2} - 2\frac{ix^2}{\beta_0 x_0^2}} \times F\left(1/2x\sqrt{2}\sqrt{\frac{i(\beta_0^2 - 1)}{\beta_0 x_0^2}}, i\right)$$

$$- \beta_0^3 x_0^2 x^3 \sqrt{\frac{i\beta_0}{x_0^2} - \frac{i}{\beta_0 x_0^2}}\sqrt{x^4\beta_0^4 - 2x^4\beta_0^2 + 4\beta_0^2 x_0^4 + x^4}$$

$$+ \beta_0 x_0^2 x^3 \sqrt{\frac{i\beta_0}{x_0^2} - \frac{i}{\beta_0 x_0^2}}\sqrt{x^4\beta_0^4 - 2x^4\beta_0^2 + 4\beta_0^2 x_0^4 + x^4}$$

$$+ \beta_0^5 x_0^2 x^5 \sqrt{\frac{i\beta_0}{x_0^2} - \frac{i}{\beta_0 x_0^2}} - 2\beta_0^3 x_0^2 x^5 \sqrt{\frac{i\beta_0}{x_0^2} - \frac{i}{\beta_0 x_0^2}}$$

$$+ 4\beta_0^3 x_0^6 x \sqrt{\frac{i\beta_0}{x_0^2} - \frac{i}{\beta_0 x_0^2}} + \beta_0 x_0^2 x^5 \sqrt{\frac{i\beta_0}{x_0^2} - \frac{i}{\beta_0 x_0^2}}$$

(34)

and

$$AD = 6c\beta_0^2 x_0^4 \sqrt{\frac{i\beta_0}{x_0^2} - \frac{i}{\beta_0 x_0^2}}\sqrt{x^4\beta_0^4 - 2x^4\beta_0^2 + 4\beta_0^2 x_0^4 + x^4}$$

(35)

where $i = \sqrt{-1}$ and

$$F(x;m) = \int_0^x \frac{1}{\sqrt{1-t^2}\sqrt{1-m^2 t^2}}\,dt$$

(36)

is the elliptic integral of the first kind, see formula 17.2.7 in [16] . Figure 2 shows the behaviour of β as function of the distance.

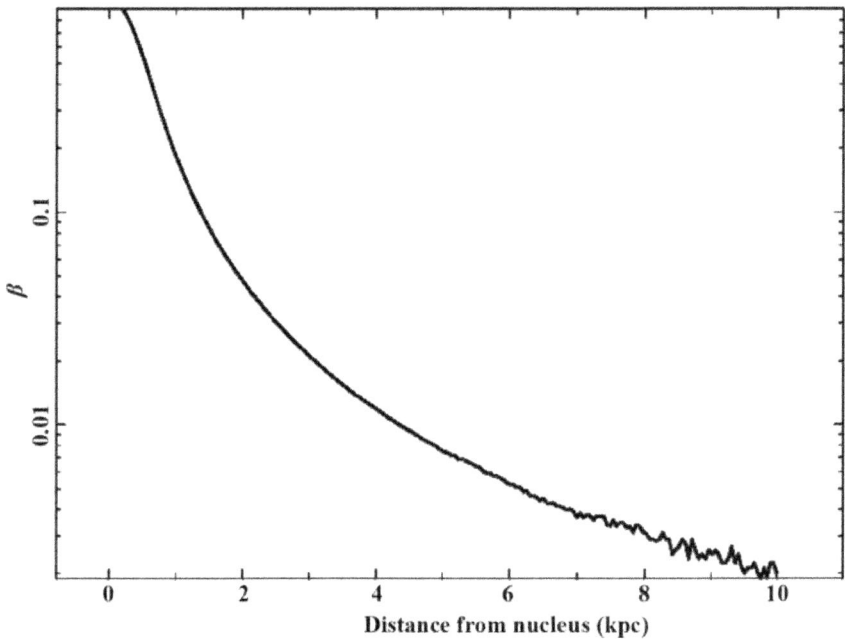

Distance from nucleus (kpc)

Figure 2: Relativistic β as a function of the distance from the nucleus when $x_0 = 200\ \text{pc}$ and $\beta_0 = 0.9$ in the case of constant density.

A numerical solution can be found by solving the following non-linear equation

$$I\left(x; \beta_0, c, x_0\right) - I\left(x_0; \beta_0, c, x_0\right) = t \tag{37}$$

and Figure 3 presents a typical comparison with the series solution.

The relativistic rate of mass flow in the case of constant density is

$$\dot{m}(x) = \frac{\rho\left(\beta_0^2 x^2 - x^2 + \sqrt{x^4 \beta_0^4 - 2x^4 \beta_0^2 + 4\beta_0^2 x_0^4 + x^4}\right) c\pi x \left(\tan\left(\alpha/2\right)\right)^2}{\sqrt{2\left(1 - \beta_0^2\right)\left(\beta_0^2 x^2 - x^2 + \sqrt{x^4 \beta_0^4 - 2x^4 \beta_0^2 + 4\beta_0^2 x_0^4 + x^4}\right)}} \tag{38}$$

Inverse Power Law Profile of Density in SR

The conservation of the relativistic energy flux in the presence of an inverse power law density profile as given by Equation (21) is

$$\rho_0 c^2 \left(\frac{d}{dt}x(t)\right)\pi(x(t))^2\left(\tan\left(\frac{\alpha}{2}\right)\right)^2\left(\frac{x_0}{x(t)}\right)^{\delta}\left(-\frac{\left(\frac{d}{dt}x(t)\right)^2}{c^2}+1\right)^{-1}$$

$$-\rho_0 c^2 v_0 \pi x_0^2\left(\tan\left(\frac{\alpha}{2}\right)\right)^2\left(-\frac{v_0^2}{c^2}+1\right)^{-1}=0.$$

(39)

This differential equation does not have an analytical solution. An expression for β as a function of the distance is

$$\beta(x)=\frac{1}{2}\frac{1}{\beta_0 x_0^2}\left(\beta_0^2 x^2\left(\frac{x_0}{x}\right)^{\delta}-x^2\left(\frac{x_0}{x}\right)^{\delta}+\sqrt{D}\right)$$

(40)

with

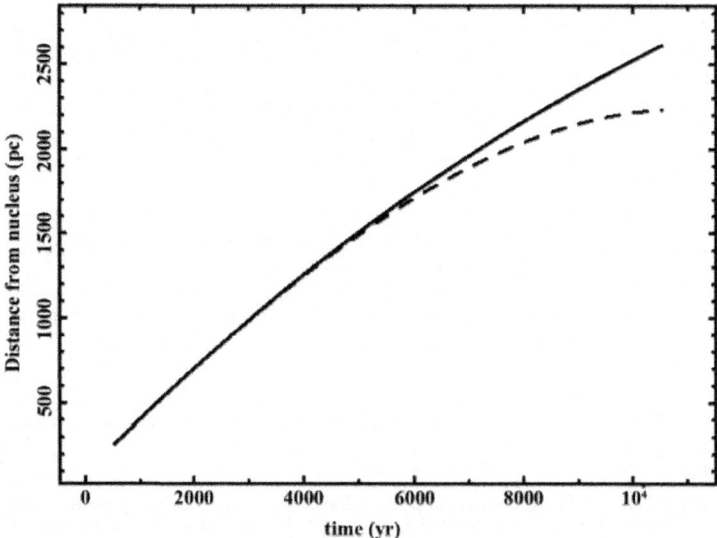

Figure 3: Non-linear relativistic solution as given by Equation (37) (full line) and series solution as given by Equation (29) (dashed line) when $x_0=100\,\mathrm{pc}$ and $\beta_0=0.999$.

$$D=\left(\left(\frac{x_0}{x}\right)^{\delta}\right)^2\beta_0^4 x^4-2\left(\left(\frac{x_0}{x}\right)^{\delta}\right)^2\beta_0^2 x^4+\left(\left(\frac{x_0}{x}\right)^{\delta}\right)^2 x^4+4\beta_0^2 x_0^4.$$

(41)

The behaviour of β as a function of the distance for different values of δ can be seen in Figure 4. A power series solution for the above differential equation (39) up to order three gives

$$a_0 = x_0$$

$$a_1 = v_0$$

$$a_2 = \frac{1}{2} \frac{v_0^2 \left(c^2\delta - \delta v_0^2 - 2c^2 + 2v_0^2 \right)}{x_0 \left(c^2 + v_0^2 \right)}.$$

(42)

Figure 5 shows a comparison between the numerical solution of (39) with the series solution.

Non-linear relativistic solution as given by Equation (39) (full line) and series solution as given by Equation (42) (dashed line) when $x_0 = 100 \, \text{pc}$ and $\beta_0 = 0.999$.

The relativistic rate of mass flow in the case of an inverse power law for the density is

$$\dot{m}(x) = \frac{\rho_0 \left(\dfrac{x_0}{x} \right)^{\delta} \left[\beta_0^2 x^2 \left(\dfrac{x_0}{x} \right)^{\delta} - x^2 \left(\dfrac{x_0}{x} \right)^{\delta} + \sqrt{D} \right] c\pi x^2 \left(\tan\left(\alpha/2 \right) \right)^2}{2\beta_0 x_0^2 \sqrt{-1/4 \dfrac{1}{\beta_0^2 x_0^4} \left(\beta_0^2 x^2 \left(\dfrac{x_0}{x} \right)^{\delta} - x^2 \left(\dfrac{x_0}{x} \right)^{\delta} + \sqrt{D} \right)^2 + 1}}$$

(43)

where ρ_0 is the density at x_0 and D was defined in Equation (41).

THE LOSSES

The previous analysis does not cover the radiative losses. The astrophysical version of the relativistic energy flux as represented by Equation (27) is

$$\frac{dE}{dt} = 1.348 \times 10^{49} \frac{n\beta_0 R_{100}^2}{1 - \beta_0^2} \frac{\text{erg}}{\text{s}}$$

(44)

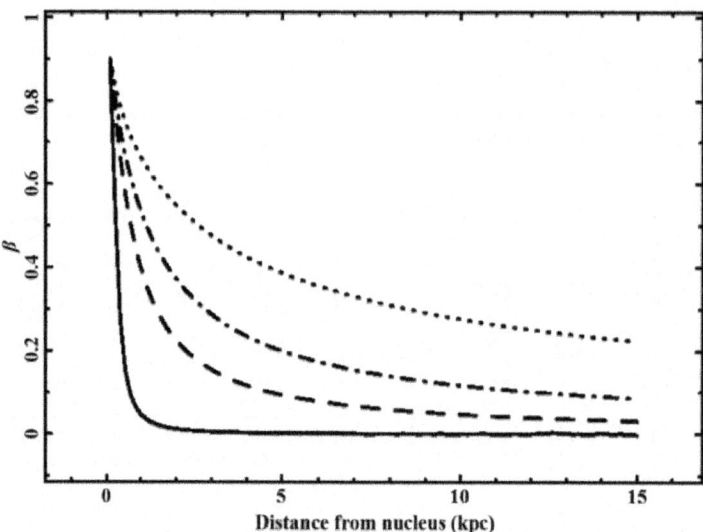

Figure 4: Relativistic β for the relativistic energy flux conservation as a function of the distance from the nucleus when $x_0 = 100\,pc$ and $\beta_0 = 0.9$: $\delta = 0$ (full line), $\delta = 1$ (dashes), $\delta = 1.2$ (dot-dash-dot-dash) and $\delta = 1.4$ (dotted).

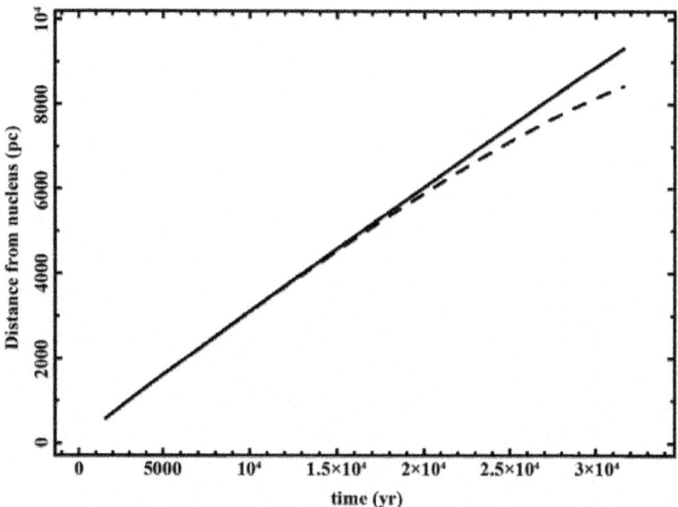

Figure 5: Non-linear relativistic solution as given by Equation (39) (full line) and series solution as given by Equation (42) (dashed line) when $x_0 = 100\,pc$ and $\beta_0 = 0.999$.

where R_{100} is the radius of the jet expressed in units of 100 pc, and n is the number density of protons expressed in particles cm^{-3}. The above luminosity is 4 - 5 orders of magnitude too high for the radio sources here considered. In order to explain this discrepancy, one model assumes that extragalactic jets are much lighter than the surroundings. The second model assumes that the observed intensity of radiation, I_ν, at a given frequency ν is a fraction of the energy flux

$$I_\nu = \varepsilon \frac{dE}{dt} \frac{\text{erg}}{\text{s}}$$

(45)

where ε represents the efficiency of conversion of the relativistic energy flux into radiation. At the moment of writing there is no exact evaluation of the efficiency of conversion. We now outline two different models for the radiative losses and a model for the magnetic field.

Losses through Recursion

In the classical case, with constant density, we can model the radiative losses through the following recursive equation obtained by modifying Equation (5)

$$\frac{1}{2}\rho v_{n+1}^3 \pi \left(x_{n+1} \tan\left(\frac{\alpha}{2}\right)\right)^2 + \varepsilon \frac{1}{2}\rho v_n^3 \pi \left(x_n \tan\left(\frac{\alpha}{2}\right)\right)^2 = \frac{1}{2}\rho v_n^3 \pi \left(x_n \tan\left(\frac{\alpha}{2}\right)\right)^2$$

(46)

where

$$x_{n+1} = x_n + v_n \Delta t.$$

(47)

Here n starts from 0, v_n is the velocity at the nth step, x_n is the position at the nth step, ε is the efficiency of conversion into radiation, α is the jet's opening angle, and Δt is the temporal step. The velocity at step $n+1$ is

$$v_{n+1} = \frac{x_n^{2/3} \sqrt[3]{1 - \varepsilon v_n}}{\left(v_n \Delta t + x_n\right)^{2/3}}.$$

(48)

Figure 6 shows the velocity as a function of the distance; $\varepsilon \approx 10^{-4}$ does not modify in an appreciable way the velocity.

In the relativistic case, with constant density, the radiative losses are modeled by a modification of Eq. (28) and the following recursive equation for the velocity at step $n+1$ is obtained

$$v_{n+1} = \frac{N_n}{D_n}$$

(49)

where

$$N_n = c^4\Delta^2 t^2 v_n^2 - c^2\Delta^2 t^2 v_n^4 + 2c^4\Delta t v_n x_n - 2c^2\Delta t v_n^3 x_n$$
$$+ c^4 x_n^2 - c^2 v_n^2 x_n^2 - c^4\sqrt{S_n}$$

$$S_n = \frac{v_n^4(v_n\Delta t + x_n)^4}{c^4} + 4\frac{v_n^2\varepsilon^2 x_n^4}{c^2} - 8\frac{v_n^2\varepsilon x_n^4}{c^2} - 2\frac{v_n^2(v_n\Delta t + x_n)^4}{c^2}$$

$$+ 4\frac{v_n^2 x_n^4}{c^2} + (v_n\Delta t + x_n)^4$$

$$D_n = 2c^2 v_n x_n^2\varepsilon - 2c^2 v_n x_n^2.$$

Figure 7 shows the relativistic velocity as a function of the distance and ε .4.2. The Parametrization of the LossesThe radiative losses can also be modeled by an "ad hoc" law for the available flux of kinetic energy, which isassumed to decrease with an inverse power law of the type $\propto\left(\frac{x_0}{x}\right)^n$. The resulting differential equation in SRwith constant density is

$$\rho c^2\left(\frac{d}{dt}x(t)\right)\pi(x(t))^2\left(\tan\left(\frac{\alpha}{2}\right)\right)^2\left(1 - \frac{\left(\frac{d}{dt}x(t)\right)^2}{c^2}\right)^{-1}$$

$$-\rho c^2 v_0\pi x_0^2\left(\tan\left(\frac{\alpha}{2}\right)\right)^2\left(1 - \frac{v_0^2}{c^2}\right)^{-1}\left(\frac{x_0}{x}\right)^n = 0.$$

$$(50)$$

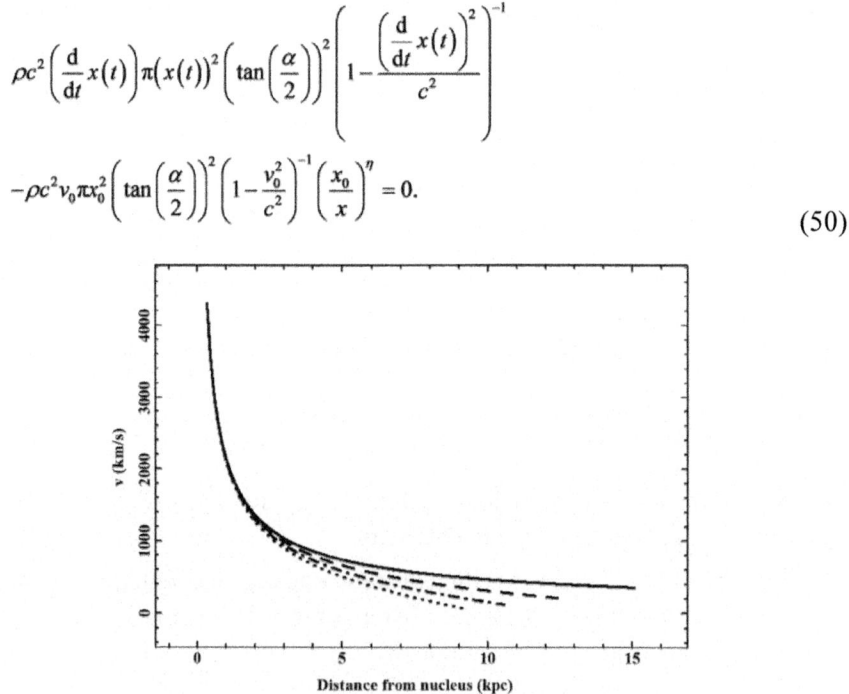

Figure 6: Classical velocity as a function of the distance from the nucleus when $x_0 = 100\,\text{pc}$, $\Delta t = 2.5\times10^4\,\text{yr}$ and $v_0 = 10,000\,\text{km}\cdot\text{s}^{-1}$: $\varepsilon = 0$(full line), $\varepsilon = 0.002$(dashes), $\varepsilon = 0.004$(dot-dash-dot-dash) and $\varepsilon = 0.006$ (dotted).

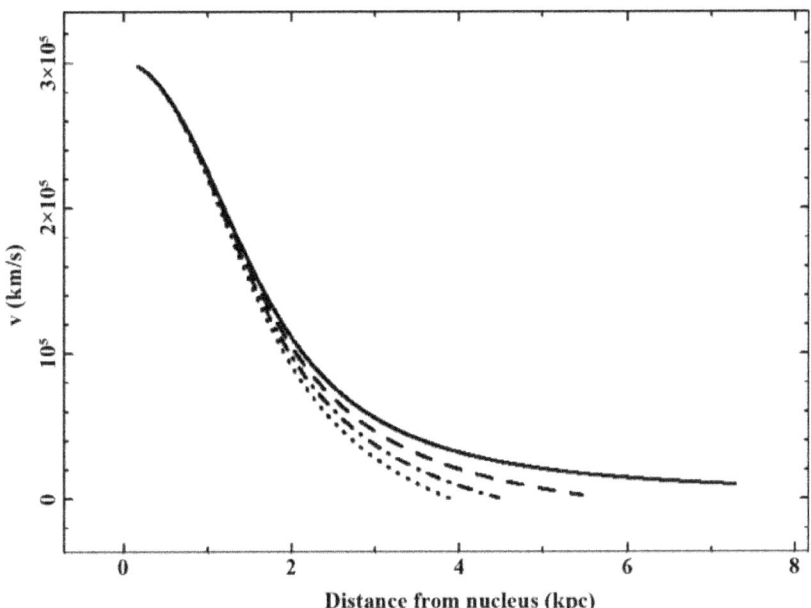

Figure 7: Relativistic velocity as a function of the distance from the nucleus when $x_0 = 100\,\mathrm{pc}$, $\Delta t = 250\,\mathrm{yr}$, and $\beta_0 = 0.999$: $\varepsilon = 0$ (full line), $\varepsilon = 0.002$ (dashes), $\varepsilon = 0.004$ (dot-dash-dot-dash) and $\varepsilon = 0.006$ (dotted).

Figure 8 shows the numerical trajectory as a function of time for different values of the exponent η: an increase in η means a lower value for the traveled distance.

The Magnetic Field

The magnetic field in CGS has an energy density of $\dfrac{B^2}{8\pi}$ where B is the magnetic field. The presence of the

magnetic field can be modeled by adding a second term for the density of energy in the rest frame of the moving fluid, see Equation (39) which models the relativistic flow of energy the in presence of an inverse power law

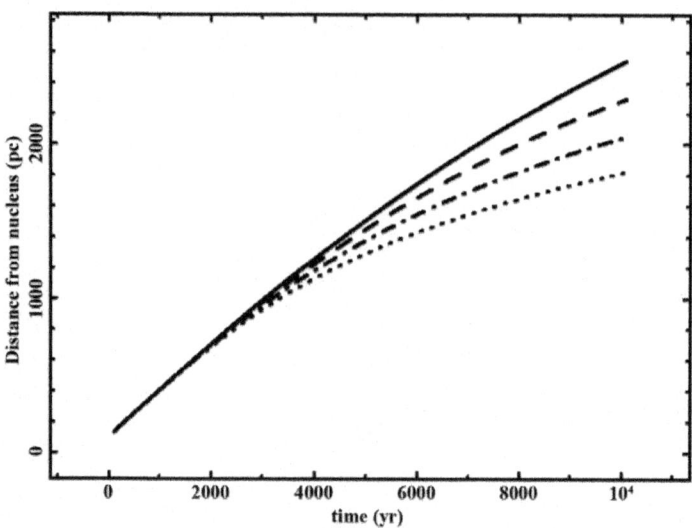

Figure 8: Relativistic distance as a function of time when $x_0 = 100\,\text{pc}$, and $\beta_0 = 0.999$; $\eta = 0$ (full line), $\eta = 0.2$ (dashes), $\eta = 0.4$ (dot-dash-dot- dash) and $\eta = 0.6$ (dotted).

$$\left(\rho_0 c^2 + \frac{B^2}{8\pi}\right)\left(\frac{d}{dt}x(t)\right)\pi\left(x(t)\right)^2\left(\tan\left(\frac{\alpha}{2}\right)\right)^2\left(\frac{x_0}{x(t)}\right)^\delta\left(-\frac{\left(\frac{d}{dt}x(t)\right)^2}{c^2}+1\right)^{-1}$$

$$-\left(\rho_0 c^2 + \frac{B_0^2}{8\pi}\right)v_0\pi x_0^2\left(\tan\left(\frac{\alpha}{2}\right)\right)^2\left(-\frac{v_0^2}{c^2}+1\right)^{-1} = 0.$$

(51)

We continue assuming a constant of proportionality between the density of energy of the magnetic field and the rest mass all along the jet

$$\frac{B(x)^2}{8\pi} \propto \rho c^2 \propto \left(\frac{x_0}{x}\right)^\delta.$$

(52)

The magnetic field as a function of the distance x is

$$B = \sqrt{B_0^2\left(\frac{x_0}{x}\right)^\delta}$$

(53)

where B_0 is the magnetic field at $x = x_0$. We assume an inverse power law spectrum for the ultrarelativistic electrons of the type

$$N(E)dE = KE^{-p}dE$$

(54)

where K is a constant and p the exponent of the inverse power law. The intensity of the synchrotron radiation has a standard expression, as given by formula (1.175) in [17],

$$I(v) \approx 0.933 \times 10^{-23} \alpha_p(p) K l H_{\perp}^{(p+1)/2} \left(\frac{6.26 \times 10^{18}}{v} \right)^{(p-1)/2} \quad erg \cdot sec^{-1} \cdot cm^{-2} \cdot Hz^{-1} \cdot rad^{-2}$$

(55)

where v is the frequency, $H_{\perp}$ is the magnetic field perpendicular to the electron's velocity, l is the dimension of the radiating region along the line of sight, and $\alpha_p(p)$ is a slowly varying function of p which is of the order of unity. As an example, $p = 2.5$ produces an intensity of the type $I(v) \propto v^{-0.75}$.

We now analyse the intensity along the centerline of the jet, which means constant radiating length. The intensity, assuming a constant p, scales as

$$I(x) = \frac{I_0 B(x)^{p/2+1/2}}{B_0^{p/2+1/2}}$$

(56)

where I_0 is the intensity at $x = x_0$ and B_0 the magnetic field at $x = x_0$. We insert Equation (53) in order to have an analytical expression for the centerline intensity

$$I(x) = B_0^{-p/2-1/2} I_0 \left(B_0^2 \left(\frac{x_0}{x} \right)^{\delta} \right)^{p/4+1/4}$$

(57)

and Figure 9 shows the theoretical synchrotron intensity as well the observed one in 3C31, seeFigure 8 in [12] . We test the goodness of fit through two standard statistical tests. The first test is the χ^2, which is computed as

$$\chi^2 = \sum_{j=1}^{n} \left(I_{obs} - I_{theo} \right)^2$$

(58)

where the index j varies from 1 to the number of available observations, n, I_{obs} is the observed intensity at position j, and I_{theo} is the observed one. A second test of the model works over different points of the jet and an observational percentage of reliability, ϵ_{obs}, is introduced

$$\epsilon_{obs} = 100 \left(1 - \frac{\sum_j |I_{obs} - I_{theo}|_j}{\sum_j I_{theo,j}} \right).$$

(59)

Another application is to the spatial evolution of the magnetic field of 3C273 as observed by VLBA in the pc region, see [18] . Figure 10 shows the observed behaviour of the magnetic field as well the theoretical evolution as represented by Equation (53).

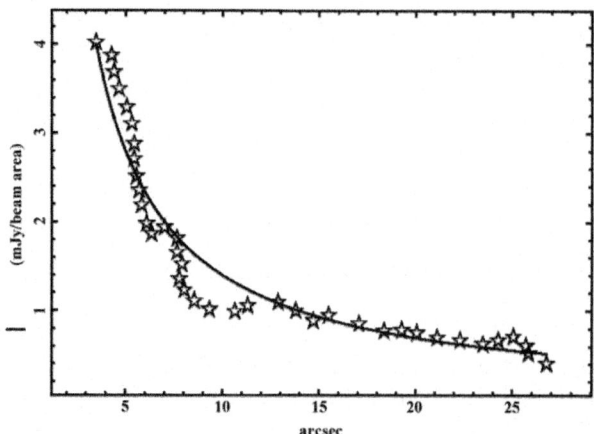

Figure 9: Intensity profile along the centerline of 3C31 when $x_0 = 3.51 \, \text{arcsec}$, $I_0 = 4$ mJy/ (beam area), $p = 2.5$, $B_0 = 10^{-4}$ gauss, $\delta = 1.15$, $\epsilon_{obs} = 87.56\%$ and $\chi^2 = 3.05$.

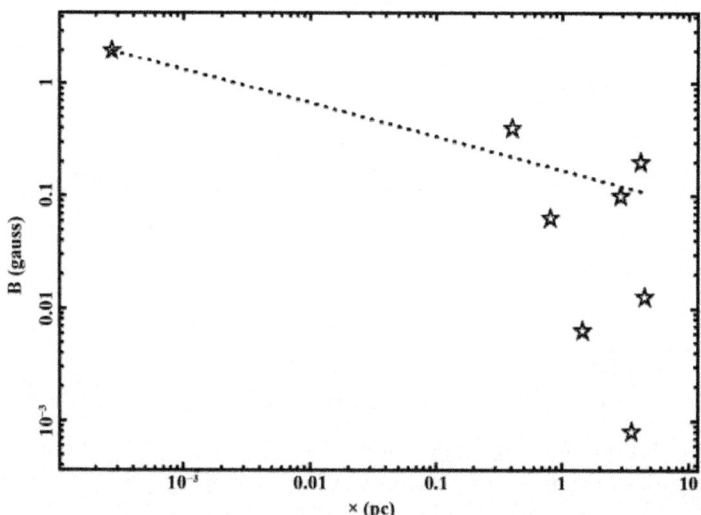

Figure 10: Observed magnetic field density of 3C273 as a function of the distance, empty stars, and theoretical curve as represented by Equation (53), dotted line, when $x_0 = 2.6 \times 10^{-4}$ pc, $B_0 = 2$ gauss, $\delta = 0.6$.

The analytical expression for the magnetic field as a function of the distance allows finding the maximum energy which can be reached in the process of acceleration of the cosmic rays in extragalactic radio-sources. The Hillas argument, see [19] , firstly introduces the relativistic ions' gyro-radius, ρ_z, expressing the energy in 10^{15} eV units (E_{15}), the magnetic field in 10^{-6} gauss (B_{-6})

$$\rho_Z = 1.08 \frac{E_{15}}{B_{-6}Z} \text{ pc}$$
(60)

where Z is the atomic number. The relativistic gyro-radius is equalized to the maximum transversal dimension of the jet, which is the diameter,

$$\rho_Z = 2x \tan\left(\frac{\alpha}{2}\right).$$
(61)

The resulting expression for the maximum energy is

$$E_{15} = 9.25 \times 10^5 x \tan\left(\frac{\alpha}{2}\right)\sqrt{B_0^2\left(\frac{x_0}{x}\right)^{\delta}}\, Z$$
(62)

where B_0 is expressed in gauss and x and x_0 in pc. Figure 11 reports the Hillas plot for 3C31 from which it is possible to say that $E_{15} = 10^6$ or $E = 10^{21}$ eV can be reached at the end of the jet when the magnetic field at $x_0 = 100$ pc is $B_0 = 0.025$ gauss.

CONCLUSIONS

Classical turbulence: We modeled the physics of turbulent jets by the conservation of the energy flux. In the case of constant density, we derived solutions for the distance and velocity as functions of time, see Equation (6) and Equation (8). In the presence of an hyperbolic profile of density, the solutions for the distance and velocity as functions of time are Equation (16) and Equation (18). The case of a density which follows an inverse power law of density is limited to the derivation of the velocity, see Equation (23). The presence of an inverse power law introduces flexibility in the results and as an example when $\delta = 2$ the rate of mass flow does not increase with x but is constant, see Equation (24).

Relativistic turbulence: The conservation of the relativistic energy flux for turbulent jets is here analysed in

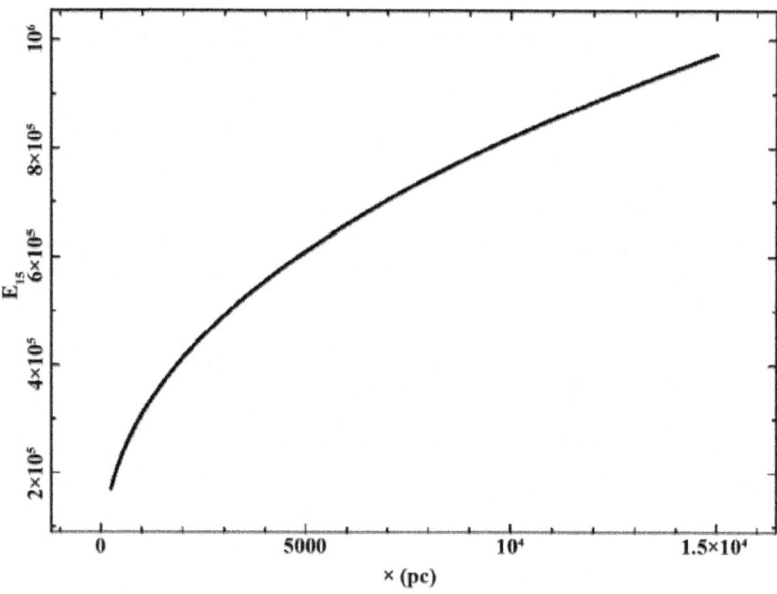

Figure 11: Maximum achievable energy, E_{15}, as a function of the distance when $x_0 = 100 \, \text{pc}$, $\beta_0 = 0.9$, $B_0 = 0.025$ gauss, $\alpha = 0.1$ and $Z = 1$.

two cases. In the first case we have a surrounding medium with constant density and the analytical result is limited to a series expansion for the solution, see Equation (29). In the second case the surrounding density decreases with a power law behaviour and the analytical result is limited to the velocity-distance relation, see Equation (40) and to a series expansion for the solution, see Equation (42).

The losses: The choice of the flux of energy as a quantity to be conserved allows a parametrization of the losses. In the first model we considered the decrease of the available classical and relativistic flux of energy through a recursive relation, see Equation (46) and Equation (49). Figure 6 andFigure 7 show the velocity as a function of the regulating parameter ε. Values of $\varepsilon < 0.001$ do not affect the jet's trajectory at the astrophy- sical distance of 15 kpc. In the second model, we fixed a law for the decrease of the available flux of relativistic energy as a function of the distance, see Equation (50) and we derived a law for the decrease of the velocity as a function of the regulating parameter η, see Figure 8.

Astrophysical applications: We modeled the behaviour of the magnetic field assuming the conservation of the magnetic flux of energy in the case of constant density, see Equation (51). The availability of an analytical expression for the magnetic field, see the theoretical Equation (53), allows finding a law for the behaviour of the intensity of the synchrotron emission, see Equation (57). The application to the measured intensity of 3C31 yields an efficiency over all the jet's length of 87.56%, see Figure 9. A test on the magnetic field of 3C273 in the pc region can be seen inFigure 10. The presence of a law for the magnetic field allows fixing the Hillas plot for the maximum energy which can reached during the process of acceleration of the cosmic rays, which in the case of 3C31 is »10^{21} eV, see the caption of Figure 11.

REFERENCES

1. Reynolds, O. (1883) An Experimental Investigation of the Circumstances Which Determine Whether the Motion of Water Shall Be Direct or Sinuous, and of the Law of Resistance in Parallel Channels. Proceedings of the Royal Society of London, 174, 935-982.

2. Reynolds, O. (1894) On the Dynamical Theory of Incompressible Viscous Fluids and the Determination of the Criterion. Proceedings of the Royal Society of London, 56, 40-45.http://dx.doi.org/10.1098/rspl.1894.0075

3. van Dyke, M. (1982) An Album of Fluid Motion. NASA STI/Recon Technical Report A, 82, 36549.

4. Goldstein, S. (1965) Modern Developments in Fluid Dynamics. Dover, New York.

5. Landau, L. (1987) Fluid Mechanics. 2nd Edition, Pergamon Press, New York.

6. Pope, S.B. (2000) Turbulent Flows. Cambridge University Press, Cambridge, UK.http://dx.doi.org/10.1017/CBO9780511840531

7. Bird, R., Stewart, W. and Lightfoot, E. (2002) Transport Phenomena. 2nd Edition, John Wiley and Sons, New York.

8. Lebedev, S.V., Suzuki-Vidal, F., Ciardi, A., Bocchi, M., Bland, S.N., Burdiak, G., Chittenden, J.P., de Grouchy, P., Hall, G.N., Harvey-Thompson, A., Marocchino, A., Swalding, G., Frank, A., Blackman, E.G. and Camenzind, M. (2011) Laboratory Simulations of Astrophysical Jets. In: Bonanno, A., de Gouveia Dal Pino, E. and Kosovichev, A.G., Eds., IAU Symposium vol. 274 of IAU Symposium, 26-35.

9. Suzuki-Vidal, F., Lebedev, S.V., Krishnan, M., Bocchi, M., Skidmore, J., Swadling, G., Harvey-Thompson, A.J., Burdiak, G., de Grouchy, P., Pickworth, L., Suttle, L., Bland, S.N., Chittenden, J.P., Hall, G.N., Khoory, E., Wilson-Elliot, K., Madden, R.E., Ciardi, A. and Frank, A. (2012) Laboratory Astrophysics Experiments Studying Hydrodynamic

and Magnetically-Driven Plasma Jets. Journal of Physics Conference Series, 370, 012002.http://dx.doi.org/10.1088/1742-6596/370/1/012002

10. De Young, D.S. (2002) The Physics of Extragalactic Radio Sources. University of Chicago Press, Chicago.

11. Mistry, D. and Dawson, J.R. (2014) Experimental Investigation of Entrainment Processes of a Turbulent Jet. 19th Australasian Fluid Mechanics Conference, Melbourne, 8-11 December 2014, 4 p.

12. Laing, R.A. and Bridle, A.H. (2002) Relativistic Models and the Jet Velocity Field in the Radio Galaxy 3C 31. MNRAS, 336, 328.

13. Palacios, A.F. (2015) The Mass-Energy Equivalence Principle in Fluid Dynamics. Journal of High Energy Physics, Gravitation and Cosmology, 1, 48-54.http://dx.doi.org/10.4236/jhepgc.2015.11005

14. Tenenbaum, M. and Pollard, H. (1963) Ordinary Differential Equations: An Elementary Textbook for Students of Mathematics, Engineering, and the Sciences. Dover Publications, New York.

15. Ince, E.L. (2012) Ordinary Differential Equations. Courier Dover Publications, New York.

16. Abramowitz, M. and Stegun, I.A. (1965) Handbook of Mathematical Functions with Formulas, Graphs, and Mathematical Tables. Dover, New York.

17. Lang, K.R. (1980) Astrophysical Formulae. 2nd Edition, Springer, New York.http://dx.doi.org/10.1007/978-3-662-21642-2

18. Savolainen, T., Wiik, K., Valtaoja, E. and Tornikoski, M. (2008) Magnetic Field Structure in the Parsec Scale Jet of 3C 273 from Multifrequency VLBA Observations. In: Rector, T.A. and De Young, D.S., Eds., Extragalactic Jets: Theory and Observation from Radio to Gamma Ray Vol. 386 of Astronomical Society of the Pacific Conference Series, 451. (Preprint)

19. Hillas, A.M. (1984) The Origin of Ultra-High-Energy Cosmic Rays. Annual Review of Astronomy and Astrophysics, 22, 425-444.

CITATION

CHAPTER 1

Beckwith, A. (2016) Gedanken Experiment Examining How Kinetic Energy Would Dominate Potential Energy, in Pre-Planckian Space-Time Physics, and Allow Us to Avoid the BICEP 2 Mistake. Journal of High Energy Physics, Gravitation and Cosmology, 2, 75-82. doi: 10.4236/jhepgc.2016.21008.

CHAPTER 2

D M South, Data Preservation and Long Term Analysis in High Energy Physics, DOI http://dx.doi.org/10.1088/1742-6596/396/6/062018.

CHAPTER 3

V. González, D. Barrientos, J. M. Blasco, F. Carrió, X. Egea and E. Sanchis (2012). Data Acquisition in Particle Physics Experiments, Data Acquisition Applications, Prof. Zdravko Karakehayov (Ed.), ISBN: 978-953-51-0713-2, InTech, DOI: 10.5772/48463.

CHAPTER 4

M. Naschie, S. Olsen, J. He, S. Nada, L. Marek-Crnjac and A. Helal, "On the Need for Fractal Logic in High Energy Quantum Physics," International Journal of Modern Nonlinear Theory and Application, Vol. 1 No. 3, 2012, pp. 84-92. doi: 10.4236/ijmnta.2012.13012.

CHAPTER 5

Xuemin Zhu and Sen Qian (2012). The Quality Management of The R&D in High Energy Physics Detector, Latest Research into Quality Control, Dr. Isin Akyar (Ed.), ISBN: 978-953-51-0868-9, InTech, DOI: 10.5772/51434.

CHAPTER 6

Dagmar Adamová and Pablo Saiz (2012). Grid Computing in High Energy Physics Experiments, Grid Computing - Technology and Applications, Widespread Coverage and New Horizons, Dr. Soha Maad (Ed.), ISBN: 978-953-51-0604-3, InTech, DOI: 10.5772/36301.

CHAPTER 7

Daniele Rinaldi, Michel Lebeau, Nicola Paone, Lorenzo Scalise and Paolo Pietroni (2011). Quality Control and Characterization of Scintillating Crystals for High Energy Physics and Medical Applications, Wide Spectra of Quality Control, Dr. Isin Akyar (Ed.), ISBN: 978-953-307-683-6, InTech, DOI: 10.5772/24010.

CHAPTER 8

Yuancun Nie, Yuanrong Lu, Xueqing Yan and Jiaer Chen (2012). Radio Frequency Quadrupole Accelerator: A High Energy and High Current Implanter, Ion Implantation, Prof. Mark Goorsky (Ed.), ISBN: 978-953-51-0634-0, InTech, DOI: 10.5772/33823.

CHAPTER 9

Beckwith, A. (2016) Gedanken Experiment for Refining the Unruh Metric Tensor Uncertainty Principle via Schwartz Shield Geometry and Planckian Space-Time with Initial Nonzero Entropy and Applying the Riemannian-Penrose Inequality and Initial Kinetic Energy for a Lower Bound to Graviton Mass (Massive Gravity). Journal of High Energy Physics, Gravitation and Cosmology, 2, 106-124. doi: 10.4236/jhepgc.2016.21012.

CHAPTER 10

Zaninetti, L. (2016) Classical and Relativistic Flux of Energy Conservation in Astrophysical Jets. Journal of High Energy Physics, Gravitation and Cosmology, 2, 41-56. doi: 10.4236/jhepgc.2016.21005.

INDEX